AC and DC Network Theory

PHYSICS AND ITS APPLICATIONS

Series Editor

E.R. Dobbs
University of London

This series of short texts on advanced topics for students, scientists and engineers will appeal to readers seeking to broaden their knowledge of the physics underlying modern technology.

Each text provides a concise review of the fundamental physics and current developments in the area, with references to treatises and the primary literature to facilitate further study. Additionally texts providing a core course in physics are included to form a ready reference collection.

The rapid pace of technological change today is based on the most recent scientific advances. This series is, therefore, particularly suitable for those engaged in research and development, who frequently require a rapid summary of another topic in physics or a new application of physical principles in their work. Many of the texts will also be suitable for final year undergraduate and postgraduate courses.

1. **Electrons in Metals and Semiconductors**
   R.G. Chambers
2. **Basic Digital Electronics**
   J.A. Strong
3. **AC and DC Network Theory**
   A.J. Pointon and H.M. Howarth

AC and DC Network Theory

A. J. Pointon

Professor, Microwave Physics,
Portsmouth Polytechnic

and

H. M. Howarth

Principal Lecturer, Physics,
Portsmouth Polytechnic

SPRINGER-SCIENCE+BUSINESS MEDIA, B.V.

First edition 1991

© 1991 A.J. Pointon and H.M. Howarth
Originally published by Chapman & Hall in 1991

Typeset in 10/12 pt Times by
Thomson Press (India) Ltd, New Delhi

British Library Cataloguing in Publication Data

Pointon, A.J.
   AC and DC network theory.
   1. Electric equipment
   I. Title   II. Howarth, H.M.   III. Series
   621.3192

   ISBN 978-0-412 38310 6   ISBN 978-94-011-3142-1 (eBook)
   DOI 10.1007/978-94-011-3142-1

Library of Congress Cataloging-in-Publication Data

Pointon, A.J., 1932–
   AC and DC network theory / A.J. Pointon and H.M. Howarth.—1st ed.
      p.   cm.—(Physics and its applications; 3)
   Includes index.

   1. Electric networks.   I. Howarth, H.M., 1929—
II. Title.   III. Series.
TK454.2.P65   1991   90-48674
621.319′2—dc20   CIP

Contents

Preface   **ix**

1 Introduction   **1**
  1.1 Electric current   1
  1.2 Electrical potential and emf   2
  1.3 Resistance and conductance   3
  1.4 Internal resistance   4
  1.5 Inductance   4
  1.6 Capacitance   5

2 Direct current theory   **6**
  2.1 Kirchhoff's laws   6
  2.2 Voltage and current generators   8
  2.3 Resistors in series and parallel   10
  2.4 Non-linear resistive elements   12
  2.5 Maximum power theorem   13

3 Capacitors, inductors and transients   **14**
  3.1 Capacitors in series and parallel   14
  3.2 Inductors in series and parallel   16
  3.3 Circuit transients   17

4 Alternating current theory   **23**
  4.1 Graphical representation of ac voltages   23
  4.2 Graphical addition of ac waveforms   25
  4.3 Algebraic addition of ac voltages and currents   25
  4.4 Phasor representation and the addition of ac voltages and currents   26
  4.5 Resistance, self-inductance and capacitance   29

4.6   Complex representation of ac voltages and
      currents—the j notation   32
4.7   Manipulation of complex impedances   34
4.8   Mutual inductance in an ac circuit   35
4.9   Kirchhoff's laws in ac circuits   36
4.10 Complex impedances in series and parallel   36
4.11 Admittance of an ac circuit   38
4.12 Root mean square values of ac quantities   38
4.13 Power in ac circuits   39
4.14 Complex power and the power triangle   42
4.15 Power usage: improvement of the power factor   43
4.16 Impedance matching   44
4.17 Complex frequency—s notation   46

5   Mesh or loop analysis and nodal analysis   48
5.1   Mesh (loop) analysis   48
5.2   Mesh and loop analysis applied to a generalized
      circuit   50
5.3   Application of mesh (loop) analysis to circuits
      containing mutual inductances   52
5.4   Input impedance of a network   54
5.5   Output impedance   55
5.6   Transfer impedance   56
5.7   Nodal analysis   57
5.8   Nodal analysis applied to a generalized circuit   59
5.9   Input, output and transfer admittances   59

6   Network theorems and transformations   63
6.1   Thévenin's theorem   63
6.2   Norton's theorem   66
6.3   Millman's theorem   69
6.4   The reciprocity theorem   72
6.5   The superposition theorem   73
6.6   The substitution theorem   74
6.7   The compensation theorem   74
6.8   The star–delta (λ–Δ) transformation   74
6.9   Transformation involving a mutual inductance   77

7   Electrical resonance   79
7.1   The series L–C–R circuit   79
7.2   Voltage magnification in a series L–C–R circuit 83
7.3   The parallel L–C–R circuit   84

7.4 Current magnification in a parallel $L-C-R$ circuit 86
7.5 Alternative definitions for resonance 86
7.6 Uses of series and parallel resonant circuits 87
7.7 Definitions of Q 88

8 Coupled circuits **90**
8.1 The generalized coupled circuit 90
8.2 The low frequency transformer 92
8.3 Resonant circuits coupled by a mutual inductance 97
8.4 Resonant circuits with direct coupling 103
8.5 Uses of coupled resonant circuits 105

9 Two-port networks **106**
9.1 Parametric representations of a two-port network 107
9.2 Parameter conversion 110
9.3 The loaded two-port network 111
9.4 Two-port networks connected in cascade 114
9.5 Characteristic impedance 116
9.6 Propagation constant 117
9.7 Symmetrical T and Π networks 118
9.8 Ladder networks 122
9.9 Filters 123
9.10 Attenuators 128
9.11 Transmission lines 129
9.12 Artificial delay lines 134

Appendix A Quantities and symbols used in the text **136**

Appendix B Exercises **139**

Appendix C Answers to exercises **154**

Index **157**

Preface

Whatever the field of human activity—domestic or scientific, work or leisure—it is likely that some knowledge of the behaviour of electrical circuits is required to keep the processes moving, the wheels turning. In many cases, a knowledge of Ohm's law may suffice. In others, an understanding of more complex relationships may be necessary.

In this book an attempt is made to provide, in a concise manner, an introduction to the main methods of treating electrical networks, whether they be carrying direct (dc) or alternating (ac) electrical currents. Clearly, the range of possible circuits is vast so that the simplifications which are demonstrated in the pages that follow are of great importance to the student. However, to gain the fullest benefit from such a concise presentation, the student must devote some time to the exercises which are provided in Appendix B.

The units used throughout the book are those of the International System (or SI). The various quantities which are introduced—such as current and potential and resistance—are summarized in Appendix A together with the symbols used to represent them, the unit associated with each quantity and the formula used to derive that unit from four fundamental or MKSA units.

ACKNOWLEDGEMENTS

It is a pleasure to thank Mrs. Pauline Dowie who organized the complex manuscript into an acceptable form, Mr. Danny Atkins who produced the diagrams, and Professor Roland Dobbs for his encouragement and many helpful suggestions.

Clearly there are many and varied sources which have been tapped in assembling this text; however, mention must be made of four books which have been of particular help to the authors.

Fewkes, J. H. and Yarwood, J. (1956). *Electricity and Magnetism and Atomic Physics*, University Tutorial Press.

Bleaney, B. I. and Bleaney, B. (1965). *Electricity and Magnetism*, Oxford University Press.

Duffin, W. J. (1973). *Electricity and Magnetism*, McGraw-Hill.

Lancaster, G. (1980). *DC and AC Circuits*, Oxford University Press.

Nevertheless, any errors or any lack of clarity which may arise from the search for conciseness must remain the responsibility of the authors.

Finally, those students wishing to extend their knowledge of network theory beyond the limits of this book are referred to

Bagguley, D M. S. (1973). *Electromagnetism and Linear Circuits*, Van Nostrand Reinhold.

HMH
AJP

1

Introduction

The theory of electrical networks or circuits has a very specific and useful purpose: it is to allow the calculation of the currents which will flow in the different components or branches of a particular network when one or more electrical signals, or sources of electrical energy, are applied to it. Of course, the motion of the electrical charges which make up the electrical currents takes place under the influence of either electrostatic or magnetic forces and will, therefore, strictly be determined by the basic laws of electricity and magnetism. Similarly, the effects which the moving charges cause as they pass along or are stored in conductors in different configurations will be governed by those same laws. However, for the purpose of electrical network theory, whether for constant (dc) or alternating (ac) currents, an attempt is made to simplify the analysis by expressing the various effects in terms of the properties of specific circuit elements such as resistors, capacitors or inductors. At the same time, the forces acting on the charges are expressed in terms of electromotive forces or electrical potentials while the movement of charges is described by reference to the electric currents.

A summary of the main features of the currents, the potentials and the circuit components utilized in network theory are given below. A fuller treatment of the concepts will be found in E. R. Dobbs, *Electricity and Magnetism*, Routledge, London, 1984.

1.1 ELECTRIC CURRENT

When a charge dQ passes a given point in time dt, a current will flow in the direction of the charge given by

$$I = \frac{dQ}{dt}$$

the convention being that the direction of current flow is the same as that of the motion of the positive charge.

The SI unit of current—the ampere, A—is the fundamental electrical unit.

The charge which passes the given point in a given time will be

$$Q = \int I \, dt$$

for which the SI unit is the *coulomb*, C, so that $1\,C = 1\,A\,s$.

1.2 ELECTRICAL POTENTIAL AND EMF

When an electrical current flows, charge moves from one position to another position where it has a lower energy. If the energy of one coulomb of charge flowing between the two positions changes by one joule, there is said to be a *potential difference* (or pd) between the two positions equal to one *volt* (1 V). One volt is thus $1\,J\,C^{-1}$. The pd between two points may be considered to be due to an electric field E acting between the points, E being the force, expressed in newtons per coulomb. The situation is illustrated in Fig. 1.1 where the field E acts at the position of the element of length ds, although not necessarily parallel to it. The pd between A and B will thus be the work done by the field acting on unit charge in moving it between those points, i.e.

$$V_{AB} = \int E \cdot ds$$

Clearly the unit of the field will be $V\,m^{-1}$.

If the potential of the earth is taken as zero and the *potentials* of the points A and B are taken to be V_A and V_B, i.e. the pds of those

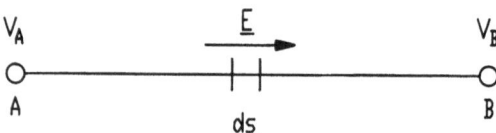

Fig. 1.1 Electric field acting between two points.

two points relative to the earth then, clearly,

$$V_{AB} = V_A - V_B$$

In a circuit which forms a closed loop, the pd which drives the current will have to be supplied to the circuit by some *electromotive force* (or emf) which is given by the integral round the loop

$$\oint E \cdot ds$$

The source of this emf may be, for example, a thermocouple or a dynamo or a voltaic cell. Such a source is said to be an *active* component in a circuit while elements such as resistors, capacitors etc., are said to be passive.

1.3 RESISTANCE AND CONDUCTANCE

Ohm's law, which is obeyed in a wide range of situations and by most, though not all, materials states that the voltage or pd across a conductor (or resistor) is proportional to the current flowing through the conductor provided its temperature remains constant. If the pd is V and the current is I then

$$V = IR$$

where the constant of proportionality, R, is called the *resistance* of the conductor for which the SI unit is the ohm, Ω, and $1\,\Omega = 1\,\mathrm{V\,A^{-1}}$. If Ohm's law had been written instead as $I = GV$, then G would have been the conductance, equal to R^{-1}, and its unit would have been the siemen, S, where $1\,\mathrm{S} = 1\,\Omega^{-1}$.

When current I flows through the resistance R, each coulomb of charge will do work $V = IR$ joules, and heating, known as *Joule heating*, will occur at the rate of VI watts. The power lost as heat is thus

$$VI = I^2 R = V^2/R \text{ watts}$$

If the temperature of the resistor changes from θ_0, at which the resistance is R_0, to a new value θ, the resistance will change, approximately, to

$$R = R_0[1 + \alpha(\theta - \theta_0)]$$

where α, the *temperature coefficient of resistance*, is generally positive. For some special alloys, e.g. manganin and constantan, $\alpha \approx 0$. For large temperature changes, a polynomial in $\theta - \theta_0$ will generally be needed to give an accurate description. In the case of semi-conductors, e.g. germanium and silicon, the resistance usually falls with temperature.

1.4 INTERNAL RESISTANCE

Most voltage generators or sources of emf will have an intrinsic or internal resistance. This situation is illustrated in Fig. 1.2 where the emf of the generator is V_0 and the internal resistance is R_0. Clearly, the pd between A and B is

$$V_{AB} = V_0 - IR_0$$

and $V_{AB} = IR$. If $R \to \infty$, $I \to 0$ so that V_0 is clearly given by V_{AB} in the open-circuit condition.

1.5 INDUCTANCE

For two coils in proximity, a current flowing in one coil will cause a linkage of magnetic flux lines to the other. If a current I_1 in the coil 1 causes a flux-linkage MI_1 to coil 2 then, by Neumann's theorem, a current I_2 in coil 2 will cause a flux-linkage MI_2 to coil 1. The quantity M is called the *mutual inductance* of the two coils. If the current in either coil changes at a rate $\mathrm{d}I/\mathrm{d}t$ then, by Faraday's law, there will be an emf generated in the other coil of $-M\,\mathrm{d}I/\mathrm{d}t$. The SI unit of mutual inductance is the *henry*, H and $1\,\mathrm{H} = 1\,\mathrm{V\,s\,A^{-1}}$ (or $1\,\mathrm{Wb\,A^{-1}}$).

For a single coil, the passage of a current will cause a magnetic flux to link to the coil itself. If unit current causes a flux linkage L,

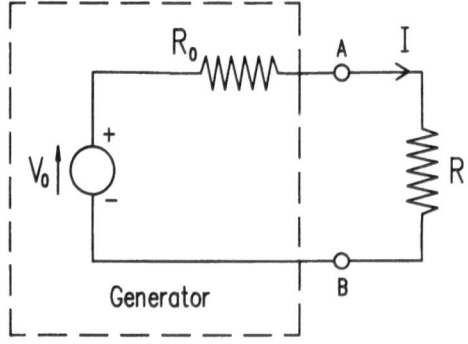

Fig. 1.2 Generator with internal resistance R_0.

then a change in the current at a rate dI/dt will generate a back emf in the coil given by

$$V_{back} = -L\frac{dI}{dt}$$

L is called the self-inductance of the coil (or *inductor*) and will also have the same unit, the henry, as M above.

Since a potential $V = -V_{back}$ will be needed to drive the current I through the coil, work will be done by the potential, equal to $I(L\,dI/dt)\,dt$ in time dt. As the current increases from zero to I, the work done will build up stored magnetic energy in the coil equal to

$$W = \int_0^I IL\,dI = \tfrac{1}{2}LI^2$$

When two coils of self-inductance L_1 and L_2 are coupled tightly so that all the flux from one circuit links with the other, it can be shown that the mutual inductance between them is

$$M = (L_1L_2)^{1/2} \tag{1.1}$$

In general there will be flux leakage and then $M = k\,(L_1L_2)^{1/2}$ where $0 \leqslant k \leqslant 1$ and k is called the *coefficient of coupling*.

1.6 CAPACITANCE

Any conductor raised to a potential V will store a charge Q which will increase proportionately with the potential. Thus it is possible to write $Q = CV$ where C is the *capacitance* of the conductor. The SI unit for capacitance is the farad, F, where $1\,F = 1\,C\,V^{-1}$. Similarly, if a *capacitor* is formed by placing two conductors in proximity, then the charge stored in the capacitor will be given by the same relation where V is the pd between the conductors and C is the capacitance of the capacitor so formed. Because the farad is a large unit, sub-multiples such as the microfarad, μF ($1\,\mu F = 10^{-6}\,F$), or the nanofarad, nF ($1\,nF = 10^{-9}\,F$), or the picofarad, pF ($1\,pF = 10^{-12}\,F$), are generally used.

Since the current is dQ/dt and $Q = CV$, the work done in time dt in charging a capacitor will be $VI\,dt = V(dQ/dt)\,dt = V\,dQ = VC\,dV$ and the energy stored in the capacitor at the potential V will be

$$W = \int_0^V VC\,dV = \tfrac{1}{2}CV^2$$

2

Direct current theory

When a constant emf is applied across a network of conducting components there will be a short period during which the current will increase from zero to a steady value. The resultant steady current is referred to as a direct current and the theory of such currents is referred to as dc theory or dc network theory. The transient variable currents which accompany any change in the parameters of the circuit, as when the emf is either connected to or removed from the circuit, are treated in the next chapter.

2.1 KIRCHHOFF'S LAWS

When a dc network is formed, it will consist of resistors connected together in various configurations together with sources of steady emf. The resulting network will possess *nodes* at junctions between two or (usually) more components. It will possess *loops* which are formed from branches of the network which combine to form closed paths, around which current may flow. It will also possess *meshes* which are simply loops containing no other loops within them.

All analysis of electrical networks is based upon Kirchhoff's laws which, when applied to the loops and nodes in the networks, allow the currents in the various branches of circuits to be calculated.

1. *Kirchhoff's current law* states that the algebraic sum of the currents at any node in a circuit is zero.
2. *Kirchhoff's voltage law* states that in any loop in a circuit the algebraic sum of the potential differences across the conducting elements is equal to the algebraic sum of the emfs acting in the loop.

The current law follows from the law of continuity which states

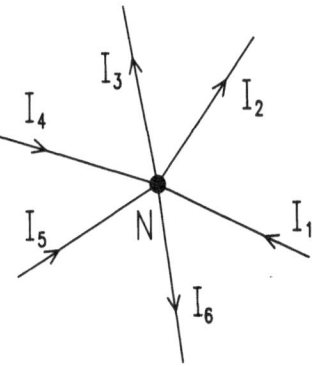

Fig. 2.1 Currents at a node N.

that there can be no continuous accumulation of charge at any point in a circuit, i.e.

$$\sum_k I_k = 0 \qquad (2.1)$$

where the sum $\sum_k$ is over all branches numbered by the integers k. Thus the net current at the node N in Fig. 2.1 can be written

$$\sum_{k=1}^{6} I_k = I_1 - I_2 - I_3 + I_4 + I_5 - I_6$$

where the direction of the current has been taken as positive when it is towards the node.

The voltage law follows from the conservation of energy since the sum of the potential differences round a loop represents the energy used in taking a unit charge around the loop, while the sum of emfs represents the energy supplied per unit charge. If, in a loop, the kth resistance R_k carries a current I_k, then the sum of the pds is $\sum_k I_k R_k$. If the jth emf is V_{0j}, then

$$\sum_k I_k R_k = \sum_j V_{0j} \qquad (2.2)$$

where the V_{0j} are added algebraically, i.e. the loop is traversed in the direction in which emfs and pds are taken as positive. (The positive direction of an emf is taken to be from its negative to its positive terminal and that of a pd is in the direction of current flow.)

An example of a network is given in Fig. 2.2 where three loops are considered. Kirchhoff's voltage law applied to this network gives:

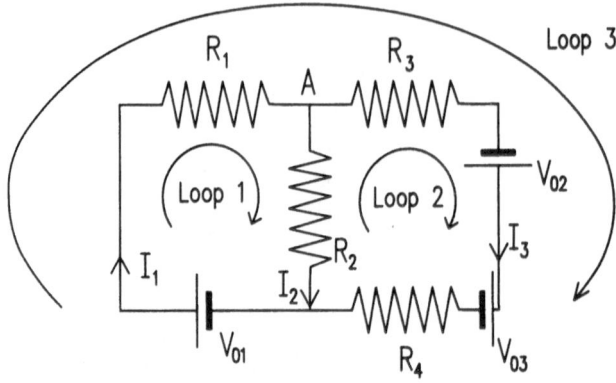

Fig. 2.2 Network with three loops (two of which are meshes).

loop 1: $\quad I_1R_1 + I_2R_2 = V_{01}$ $\hspace{3cm}$ (2.3(a))

loop 2: $\quad I_3R_3 + I_3R_4 - I_2R_2 = V_{02} - V_{03}$ $\hspace{1cm}$ (2.3(b))

loop 3: $\quad I_1R_1 + I_3R_3 + I_3R_4 = V_{01} + V_{02} - V_{03}$ $\hspace{0.4cm}$ (2.3(c))

It will be noted that, at the node A, from (2.1), $I_1 = I_2 + I_3$. This does not mean that there are four equations and only three unknowns, since (2.3(c)) is a linear combination of (2.3(a)) and (2.3(b)).

2.2 VOLTAGE AND CURRENT GENERATORS

In practice, the 'supply' to a circuit may be considered to be a generator of either voltage or current.

2.2.1 Voltage generator

The conventional representation of a voltage generator is indicated in Fig. 2.3 where V_0 is a constant emf which is in series with an internal resistance R_0. If the output terminals A and B supply an external network with a current I, the voltage which appears between A and B will be, from Kirchhoff's voltage law,

$$V = V_0 - IR_0 \hspace{3cm} (2.4)$$

2.2.2 Current generator

The formalized current generator is shown in Fig. 2.4 where the constant current source giving I_0 is in parallel with an internal

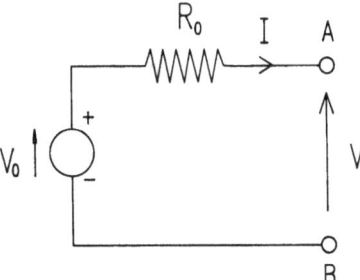

Fig. 2.3 Voltage generator.

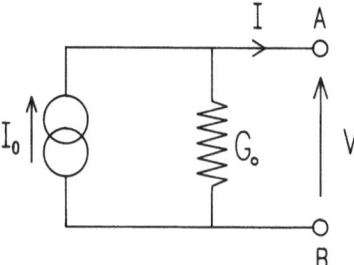

Fig. 2.4 Current generator.

conductance, G_0. If such a generator delivers a current I to a load across AB, then the voltage across the load, V, will be given by Kirchhoff's current law as

$$I = I_0 - G_0 V \tag{2.5(a)}$$

or

$$V = (I_0 - I)/G_0 \tag{2.5(b)}$$

2.2.3 The equivalence of the voltage and current generators

The generators described in Figs 2.3 and 2.4 will be equivalent if, for the same load, they deliver the same current I for the same voltage V between the terminals A and B.

If the load is infinite (i.e. open circuit), $I = 0$ and, from (2.4), $V = V_0$. Then, from (2.5(a)), $G_0 = I_0/V_0$. If the load is zero (i.e. short circuit), $V = 0$ and, from (2.5), $I = I_0$. Then, from (2.4), $R_0 = V_0/I_0$. Thus the two generators will be equivalent if

$$G_0 = 1/R_0 \tag{2.6}$$

and

$$I_0 = V_0/R_0 \tag{2.7}$$

It follows that *any* supply to a network may be considered as either a voltage generator or an equivalent current generator.

(The sources discussed above are referred to as *constant* or *independent* since the voltage V_0 or current I_0 are unaffected by changes in the network placed across the terminals. However, some sources are *dependent* because V_0 or I_0 can vary as parameters of the external network are varied.)

2.3 RESISTORS IN SERIES AND PARALLEL

In any network it is obviously advantageous to be able to replace a number of resistors by a single equivalent resistor. In order to achieve this, it is necessary first to know how to determine the resistor which is equivalent to a combination of resistors which are either in parallel or in series. Extension of this process will then enable most combinations of resistors in a network to be reduced to a single equivalent resistor.

2.3.1 Resistors in series

Suppose that a current I flows through a series of n resistors $R_1, R_2, R_3, \ldots, R_n$ in turn as shown in Fig. 2.5. The total potential drop V will, according to Kirchhoff's voltage law, be

$$V = IR_1 + IR_2 + IR_3 + \cdots + IR_n \equiv IR \qquad (2.8)$$

where R is the equivalent resistance. Then

$$R = R_1 + R_2 + R_3 + \cdots + R_n = \sum_{k=1}^{n} R_k \qquad (2.9)$$

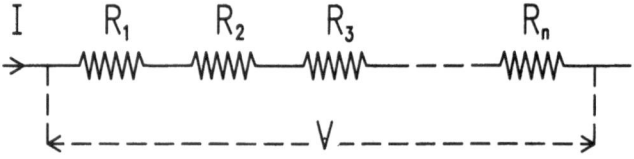

Fig. 2.5 Resistors in series carrying a common current I.

2.3.2 Resistors in parallel

If a set of n resistors are connected in parallel as shown in Fig. 2.6, the total current I is given by Kirchhoff's law as

$$I = I_1 + I_2 + I_3 + \cdots + I_n = \sum_{k=1}^{n} I_k \tag{2.10}$$

But, by Kirchhoff's voltage law, the potential difference between the nodes A and B must be the same for each resistor. Thus

$$V = I_1 R_1 = I_2 R_2 = I_3 R_3 = \cdots = I_k R_k = \cdots = I_n R_n$$

Thus

$$I_1 = V/R_1, \, I_2 = V/R_2, \, I_3 = V/R_3, \ldots, I_k = V/R_k, \ldots, I_n = V/R_n \tag{2.11}$$

From (2.10) and (2.11),

$$I = \sum_{k=1}^{n} V/R_k \equiv V/R \tag{2.12}$$

where R is the single equivalent resistor. Thus

$$1/R = 1/R_1 + 1/R_2 + 1/R_3 + \cdots + 1/R_k + \cdots + 1/R_n$$

$$= \sum_{k=1}^{n} 1/R_k \tag{2.13}$$

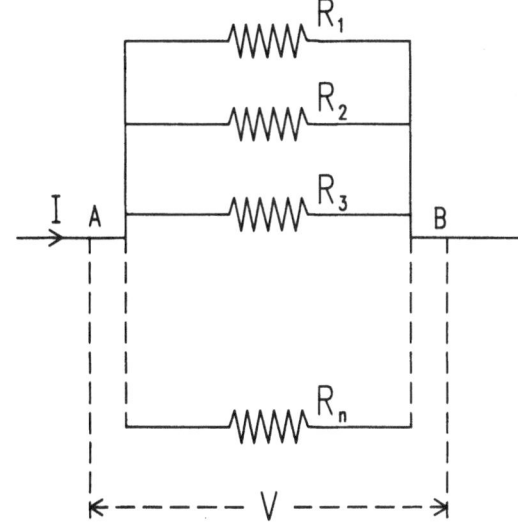

Fig. 2.6 Resistors in parallel with a common pd V.

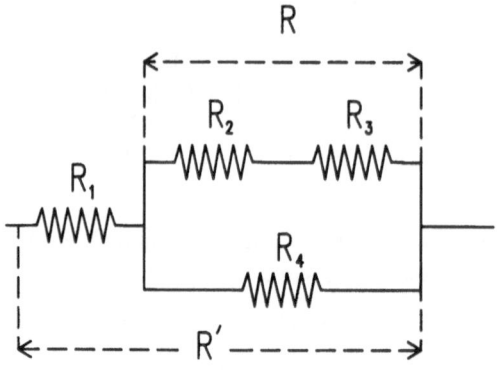

Fig. 2.7 Combination of resistors.

2.3.3 Combinations of resistors

As a simple example, Fig. 2.7 shows how a combination of resistors may be replaced by a single resistor. Since one arm of the parallel combination is equivalent to a single resistor $R_2 + R_3$, the parallel combination is thus equivalent to a resistance R where

$$\frac{1}{R} = \frac{1}{R_4} + \frac{1}{R_2 + R_3}$$

and the combination is equivalent to

$$R' = R_1 + R = R_1 + \frac{R_4(R_2 + R_3)}{R_2 + R_3 + R_4}$$

2.4 NON-LINEAR RESISTIVE ELEMENTS

There are resistive elements which do not give the linear relationship of $I = GV$ where $G = R^{-1}$ is a constant. These elements are referred to as *non-linear*. However, their current voltage relationship can often be expressed analytically. For example,

$$\textit{varistor:} \quad I = AV^n$$

$$\textit{semiconductor p–n junction:} \quad I = I_0\left[\exp\left(\frac{eV}{kT}\right) - 1\right]$$

$$\textit{Vacuum diode for } I \ll I_{SAT}: \quad I = AV^{3/2}$$

$$\textit{General:} \quad I = a + bV + cV^2$$

2.5 MAXIMUM POWER THEOREM

As the resistive load placed across a voltage generator is varied, there will be a value at which the power dissipated in the load is a maximum. The current in the variable load R shown in Fig. 2.8 is given by $I = V_0/(R + R_0)$ and the power dissipated in the load is

$$P = I^2 R = V_0^2 R/(R_0 + R)^2 \qquad (2.14)$$

Differentiating wrt R gives

$$dP/dR = V_0^2 [(R_0 + R)^2 - 2R(R_0 + R)](R_0 + R)^{-4}$$

The condition for a maximum, with $dP/dR = 0$, is

$$R = R_0 \qquad (2.15)$$

The maximum power in the load is given by

$$P_{max} = V_0/4R_0 \qquad (2.16)$$

and the load is then said to be *matched* to the generator. The power given by (2.16) is sometimes referred to as the 'available power'. It increases as the internal resistance decreases.

It is possible to state a 'maximum power theorem' for the dc case:

the power dissipated in a load by a given generator will be a maximum when the resistance of the load is equal to the internal resistance of the generator.

This theorem will apply to sources of emf such as batteries, thermocouples, dynamos and photodiodes.

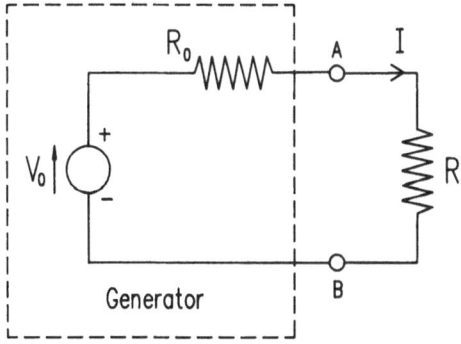

Fig. 2.8 Generator with constant voltage, V_0, internal impedance, R_0, and variable load resistor, R.

3

Capacitors, inductors and transients

This chapter is primarily concerned with the transient, variable currents which occur whenever a steady emf is suddenly applied to or removed from a circuit which, in addition to resistors, also contains capacitors and/or inductors. However, before discussing these transient effects, it is thought useful to analyse the effect of combinations of capacitors and of inductors as for resistors in the previous chapter. (A good reference for simple differential equations is C. W. Evans, *Engineering Mathematics*, Chapman and Hall, London, 1989.)

3.1 CAPACITORS IN SERIES AND PARALLEL

3.1.1 Capacitors in series

If a set of capacitors $C_1, C_2, \ldots, C_k, \ldots, C_n$ are connected in series as shown in Fig. 3.1. and a voltage V is applied across them, then the sum of the potentials across the individual capacitors will be equal

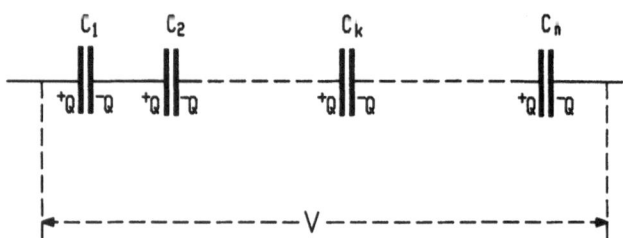

Fig. 3.1 Capacitors connected in series.

to the applied voltage V. Each capacitor must carry the same charge, Q, and hence the potential across kth capacitor is $V_k = Q/C_k$ so that

$$V = \sum_{k=1}^{n} V_k = \sum_{k=1}^{n} Q/C_k \equiv Q/C \tag{3.1}$$

where C is the single equivalent capacitor. Thus (3.1) gives

$$\frac{1}{C} = \frac{1}{C_1} + \frac{1}{C_2} + \cdots + \frac{1}{C_k} + \cdots + \frac{1}{C_n} = \sum_{k=1}^{n} \frac{1}{C_k} \tag{3.2}$$

3.1.2 Capacitors in parallel

When a set of n capacitors are connected in parallel as in Fig. 3.2, then each has the same potential V and the charge on the kth capacitor will be $Q_k = VC_k$. The total charge will be

$$Q = \sum_{k=1}^{n} Q_k = \sum_{k=1}^{n} VC_k \equiv VC \tag{3.3}$$

where C is the single equivalent capacitor. Thus (3.3) gives

$$C = C_1 + C_2 + \cdots + C_k + \cdots + C_n = \sum_{k=1}^{n} C_k \tag{3.4}$$

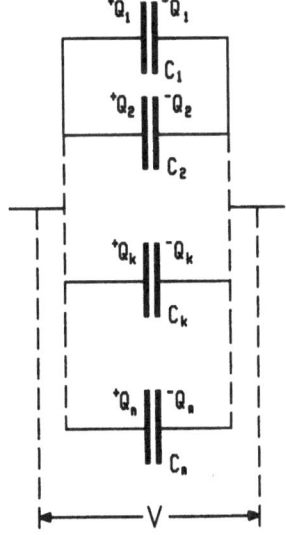

Fig. 3.2 Capacitors in parallel.

3.2 INDUCTORS IN SERIES AND PARALLEL

3.2.1 Inductors in series

When a set of n inductors is connected in series as shown in Fig. 3.3, and there is no mutual inductance between them, then each will carry the same current, I, and be subject to the same rate of change in that current, dI/dt. The potential induced in the kth inductor will be $-L_k dI/dt$ so that a potential $+L_k dI/dt$ will be required to overcome this potential. The total potential across the n inductors will thus be

$$V = \sum_{k=1}^{n} L_k \frac{dI}{dt} \equiv L \frac{dI}{dt} \tag{3.5}$$

where L is the inductance of the single equivalent inductor. Hence, from (3.5),

$$L = L_1 + L_2 + \cdots + L_k + \cdots + L_n = \sum_{k=1}^{n} L_k \tag{3.6}$$

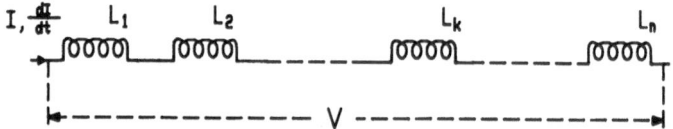

Fig. 3.3 Inductors in series.

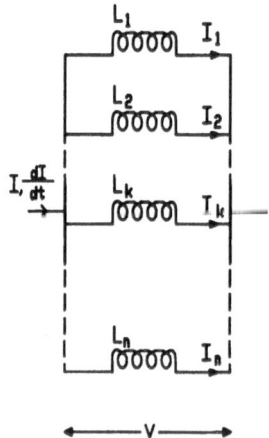

Fig. 3.4 Inductors in parallel.

3.2.2 Inductors in parallel

A set of n inductors in parallel, as shown in Fig. 3.4, and having no mutual inductance between them will each have the same potential difference across it. Thus

$$V = L_1 \frac{dI_1}{dt} = L_2 \frac{dI_2}{dt} = \cdots = L_k \frac{dI_k}{dt} = \cdots = L_n \frac{dI_n}{dt} \qquad (3.7)$$

But

$$\frac{dI}{dt} = \sum_{k=1}^{n} \frac{dI_k}{dt} = \frac{dI_1}{dt} + \frac{dI_2}{dt} + \cdots + \frac{dI_k}{dt} + \cdots + \frac{dI_n}{dt} \qquad (3.8)$$

and, combining (3.7) and (3.8),

$$\frac{dI}{dt} = \sum_{k=1}^{n} V/L_k \equiv V/L \qquad (3.9)$$

where L is the inductance of the single equivalent inductor. Equation (3.9) gives

$$\frac{1}{L} = \frac{1}{L_1} + \frac{1}{L_2} + \cdots + \frac{1}{L_k} + \cdots + \frac{1}{L_n} = \sum_{k=1}^{n} 1/L_k \qquad (3.10)$$

3.3 CIRCUIT TRANSIENTS

When a constant voltage dc source is connected to, or removed from, a network, there will be a transient period before a steady state is reached. Three main cases are considered: the R–L circuit, the R–C circuit and the R–C–L circuit.

3.3.1 R–L Circuit

Consider that there is a constant voltage V across a resistance and inductance in series as shown in Fig. 3.5. Then

$$V = RI + L \frac{dI}{dt} \qquad (3.11)$$

The general solution to (3.11) if V is constant is

$$I = (V - A e^{-Rt/L})/R \qquad (3.12)$$

where A is a constant of integration. If the voltage V is suddenly

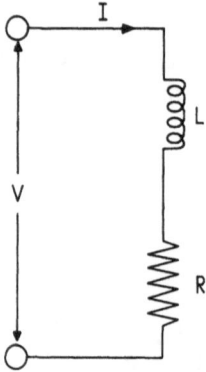

Fig. 3.5 *R* and *L* with constant voltage applied.

applied at time $t = 0$ when $I = 0$, then $A = V$ and

$$I = \frac{V}{R}(1 - e^{-Rt/L})$$ (3.13)

Similarly, if the voltage is suddenly reduced to zero at time $t = 0$ when the instantaneous current $I = I_0$, then $A = -I_0 R$ and

$$I = I_0 e^{-Rt/L}$$ (3.14)

The quantity L/R has the dimensions of time and is known as the *time constant* or, τ, of the circuit; it is the time in which the current decays from I_0 to I_0/e. The forms of the transients of (3.13) and (3.14) are shown in Figs 3.6(a) and 3.6(b) respectively.

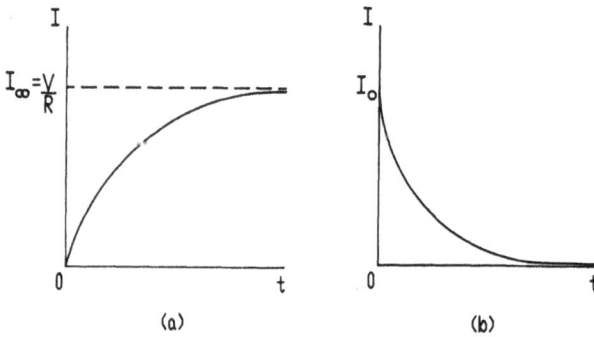

Fig. 3.6 Transients for an *R–L* circuit. (a) *V* switched on at $t = 0$. (b) *V* switched to zero at $t = 0$.

3.3.2 *R–C* circuit

When a constant voltage is applied across a resistance and capacitance in series as shown in Fig. 3.7, the current will be $I = dQ/dt$ and the charge Q on the capacitor must satisfy the differential equation

$$V = R\frac{dQ}{dt} + \frac{Q}{C} \tag{3.15}$$

Equation (3.15) has the general solution

$$Q = VC - Be^{-t/RC} \tag{3.16}$$

where B is a constant of integration.

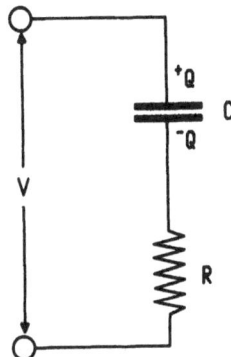

Fig. 3.7 *R* and *C* with constant voltage applied.

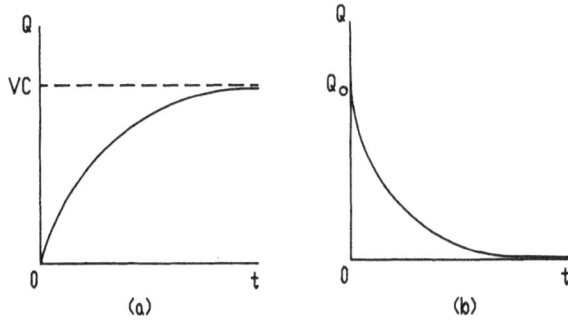

Fig. 3.8 Transients for an *R–C* circuit. (a) *V* switched on at $t = 0$. (b) *V* switched to zero at $t = 0$.

If a voltage V is suddenly applied at $t = 0$ when $Q = 0$, then $B = VC$ and

$$Q = VC(1 - e^{-t/RC}) \qquad (3.17)$$

If V is suddenly reduced to zero at $t = 0$ when the charge on the capacitor is Q_0, then $B = -Q_0$, and (3.16) gives

$$Q = Q_0 e^{-t/RC} \qquad (3.18)$$

Here $RC = \tau$, the time constant of the circuit in which the charge Q decreases from Q_0 to $Q_0 e^{-1}$. The form of Q is shown for each case in Fig. 3.8. In both cases, the current flowing can be calculated by differentiating the corresponding equation for Q with respect to time.

3.3.3 *R–C–L* circuit

When a steady voltage V is applied across an R–C–L combination as shown in Fig. 3.9, the charge will have to satisfy the differential equation

$$V = L\frac{d^2Q}{dt^2} + R\frac{dQ}{dt} + \frac{Q}{C} \qquad (3.19)$$

The general solution of (3.19) is

$$Q = VC + A_+ e^{S_+ t} + A_- e^{S_- t} \qquad (3.20)$$

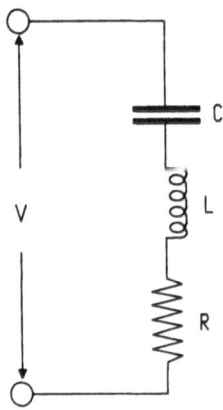

Fig. 3.9 *R–C–L* circuit with constant voltage applied.

where A_+ and A_- are constants of integration and

$$S_\pm = [-R \pm (R^2 - 4L/C)^{1/2}]/2L \qquad (3.21)$$

(When $C \to \infty$ (i.e. no capacitor), $S_+ = 0$, which requires $A_+ = 0$, and $S_- = -R/L$ as for an R-L combination. When $L \to 0$ (i.e. no inductor), $S_- = -\infty$ and $S_+ = -1/RC$ as for an R-C combination.)

By writing $m = \sqrt{R^2/4L^2 - 1/LC}$ and $b = R/2L$ in (3.21) $S_\pm = -b \pm m$ and (3.20) becomes

$$Q = VC + e^{-bt}(A_+ e^{mt} + A_- e^{-mt}) \qquad (3.22(a))$$

and

$$I = \frac{dQ}{dt} = e^{-bt}[A_+(m-b)e^{mt} - A_-(m+b)e^{-mt}] \qquad (3.22(b))$$

Only the case where the voltage is suddenly reduced to zero at $t = 0$ will be considered in more detail. Suppose that the charge on the capacitor at that instant is $Q = Q_0$. The initial current I will be zero. Inserting these boundary conditions into (3.22(a)) and (3.22(b)) gives

$$Q = Q_0 e^{-bt}[(b+m)e^{mt}/2m - (b-m)e^{-mt}/2m] \qquad (3.23)$$

and

$$I = -(Q_0/2m)(b^2 - m^2)e^{-bt}(e^{mt} - e^{-mt})$$
$$= -(Q_0/m)(b^2 - m^2)e^{-bt}\sinh mt \qquad (3.24)$$

where the negative sign occurs because the current flow is opposite to the potential which produced Q_0. The current variation can be considered under three conditions.

1. $m^2 > 0$, m real, 'overdamping'. In the case where $R^2/4L^2 > 1/LC$, m is real but less than b and the current variation is as shown in Fig. 3.10. The tail decays as $e^{-(b-m)t}$. The time, t_m, at which the maximum current occurs is found by differentiating I to give

$$t_m = \frac{1}{2m} \log_e \left(\frac{b+m}{b-m} \right) \qquad (3.25)$$

2. $m^2 < 0$, $m = j\omega$, 'underdamping'. When $R^2/4L^2 < 1/LC$, m will be imaginary and (3.24) becomes oscillatory.

$$I = -Q_0 \left(\frac{b^2 + \omega^2}{\omega} \right) e^{-bt} \sin \omega t \qquad (3.26)$$

which represents a decaying sine wave as shown in Fig. 3.11 with $\omega = \sqrt{1/LC - R^2/4L^2}$.

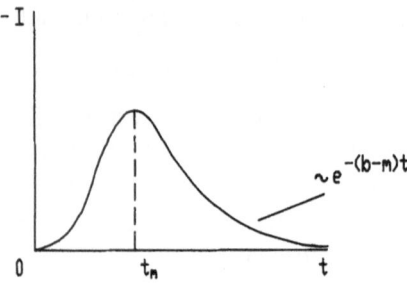

Fig. 3.10 Variation of current in an R–L–C circuit when the constant voltage is removed at $t = 0$.

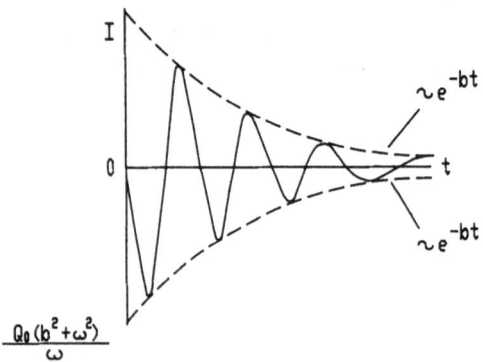

Fig. 3.11 Oscillatory current for underdamping.

3. $m = 0$, *'critical damping'*. Consider the condition that $R^2/4L^2 \rightarrow 1/LC$ (i.e., $m \rightarrow 0$); then, taking $e^{\pm mt} \rightarrow 1 \pm mt$ and substituting in (3.23),

$$Q = Q_0 e^{-bt}(1 + bt)$$

which, on differentiating wrt time, gives

$$I = - Q_0 b^2 t e^{-bt} \tag{3.27}$$

The time to reach the maximum current in this case is obtained by differentiating (3.27) to give $t_m = b^{-1}$ which is the minimum value which t_m can have for a given value of R/L. The circuit is said to be *critically damped* because, if the damping (i.e. the resistance) were further reduced, the system—or circuit—would oscillate.

4

Alternating current theory

Alternating current, or ac, theory is concerned with the mathematical analysis of the steady-state behaviour of electrical circuits in which the currents and voltages vary periodically with time. The analysis is simplified by considering only sinusoidal variations, an approach which is not restrictive since any general periodic waveform can be represented as a sum of such quantities, i.e. a Fourier series. In this chapter, it is shown how the sinusoidal waveforms can be represented both graphically and mathematically and how, in consequence, the effect of various circuit elements can be expressed in terms of generalized impedances.

4.1 GRAPHICAL REPRESENTATION OF AC VOLTAGES

An ac voltage which varies sinusoidally with time can be represented by the equation

$$V = V_0 \sin \omega t \tag{4.1}$$

and can be represented graphically (Fig. 4.1(a)), where V is the *instantaneous voltage* or *potential* at time t, V_0 is the *peak* or *maximum value* or *amplitude* of the voltage, and ω is the *angular frequency* or *pulsatance* of the wave.

The *period T*, the time for one complete cycle such that $\omega T = 2\pi$, is given by

$$T = 2\pi/\omega \tag{4.2}$$

The number of cycles or periods occurring in unit time is the

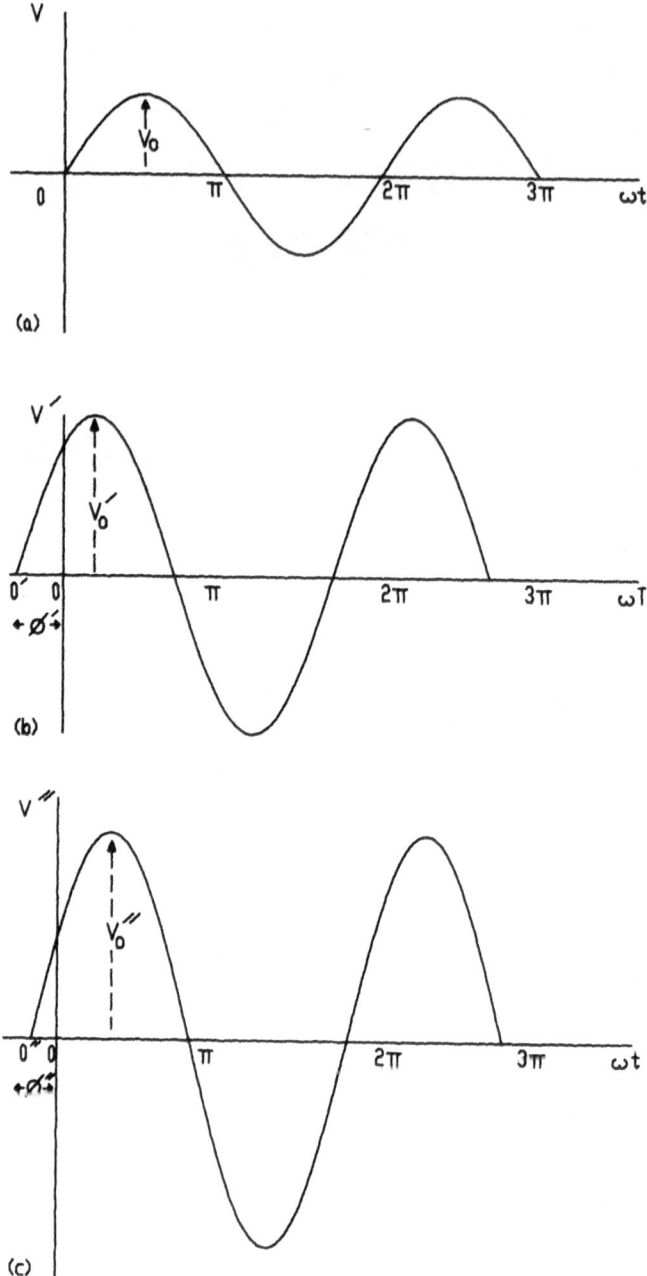

Fig. 4.1 (a) Representation of $V = V_0 \sin \omega t$. (b) Representation of V' $V'_0 \sin(\omega t + \phi')$. (c) Representation of $V'' = V + V' = V''_0 \sin(\omega t + \phi'')$.

frequency of the wave, f, and is given by

$$f = T^{-1} = \omega/2\pi \qquad (4.3)$$

In SI units, pulsatance ω is measured in *radians per second* or *rad s*$^{-1}$ and the period T is quoted in *seconds* or, if appropriate, submultiples of a second, e.g. milliseconds (ms) $\equiv 10^{-3}$ s, microsecond (μs) $\equiv 10^{-6}$ s and nanoseconds (ns) $\equiv 10^{-9}$ s. Frequency f, which has dimensions s^{-1}, is quoted in *hertz* (Hz) or in multiples such as kilohertz (kHz) $\equiv 10^{3}$ Hz, megahertz (MHz) $\equiv 10^{6}$ Hz and gigahertz (GHz) $\equiv 10^{9}$ Hz.

A second voltage V', having the same frequency as V but a different phase and a peak value V'_0 (Fig. 4.1(b)), can be represented mathematically by the equation

$$V' = V'_0 \sin(\omega t + \phi') \qquad (4.4)$$

where ϕ' is the *phase difference* between V and V'. The curve for V' is said to *lead* that for V by an angle ϕ', the lead being expressed by the positive value of ϕ' in (4.4). Conversely, V is said to *lag behind*, or simple *lags* V' by ϕ'. The phase angle is normally expressed in degrees, or multiples or fractions of π.

4.2 GRAPHICAL ADDITION OF AC WAVEFORMS

If the two voltages V and V' are applied simultaneously and in series to a circuit, the resultant voltage V'' is the instantaneous sum of V and V'. The result of adding the two curves represented in Figs 4.1(a) and 4.1(b) is shown in Fig. 4.1(c) and it can be seen that V'' (i) has a peak value V''_0, (ii) has the same frequency as its two components, V and V', and (iii) leads V by the angle ϕ'' corresponding to the separation $0''0$. Consequently, the resultant voltage V'' can be expressed as

$$V'' = V' \sin(\omega t + \phi'') \qquad (4.5)$$

This method of obtaining the resultant V'' of two voltages V and V' is clearly tedious, and is likely to be inaccurate unless the addition is made from digitized waveforms on a computer.

4.3 ALGEBRAIC ADDITION OF AC VOLTAGES AND CURRENTS

The instantaneous values of V, V' and V'' in Fig. 4.1 at a time t are related by $V'' = V + V'$ or

$$V'' = V_0 \sin \omega t + V'_0 \sin (\omega t + \phi')$$

The expansion $\sin(\omega t + \phi') = \sin \omega t \cos \phi' + \cos \omega t \sin \phi'$ gives

$$V'' = (V_0 + V'_0 \cos \phi') \sin \omega t + (V'_0 \sin \phi') \cos \omega t$$

Using the well-known result that $(a \sin \theta + b \cos \theta) = (a^2 + b^2)^{1/2} \times \sin (\theta + \delta)$ where $\delta = \tan^{-1} (b/a)$, it is possible to write

$$V'' = V''_0 \sin (\omega t + \phi'') \tag{4.6}$$

where

$$V''_0 = (V_0^2 + V'^2_0 + 2V_0 V'_0 \cos \phi')^{1/2} \tag{4.7}$$

and

$$\phi'' = \tan^{-1} \left(\frac{V'_0 \sin \phi'}{V_0 + V'_0 \cos \phi'} \right) \tag{4.8}$$

This method, although more accurate and simpler than the graphical summation, is only true when the waveforms are precisely sinusoidal—a rare occurrence in practice.

If two sinusoidal currents $I_0 \sin \omega t$ and $I'_0 \sin(\omega t + \phi')$ were to be added together, the resultant I'' would have a form corresponding to (4.6) to (4.8) with the peak currents I_0 and I'_0 replacing V_0 and V'_0.

4.4 PHASOR REPRESENTATION AND THE ADDITION OF AC VOLTAGES AND CURRENTS

It is well known that a sinusoidal motion will be generated by the projection of a vector rotating about a point with constant angular frequency. Such a vector 0A, of length V_0, is shown in Fig. 4.2 rotating at angular frequency ω in an anticlockwise sense: the projection OP can be taken to represent a voltage $V = V_0 \sin \omega t$. Expressed in this way, the vector 0A is referred to as a *phasor*.

If the two voltages $V = V_0 \sin \omega t$ and $V' = V'_0 \sin (\omega t + \phi')$ are both represented by phasors (Fig. 4.3), then the voltage V'' is represented by the phasor which is the instantaneous vector sum of V and V' and is obtained by completing the parallelogram 0ACB. Since V and V' have the same angular frequency, ω, so will their resultant V''. The *rotating phasor diagram* (Fig. 4.3) can be applied to both voltages and currents.

When each phasor has the same frequency, only the resultant amplitude and phase are of interest, and these may be obtained from

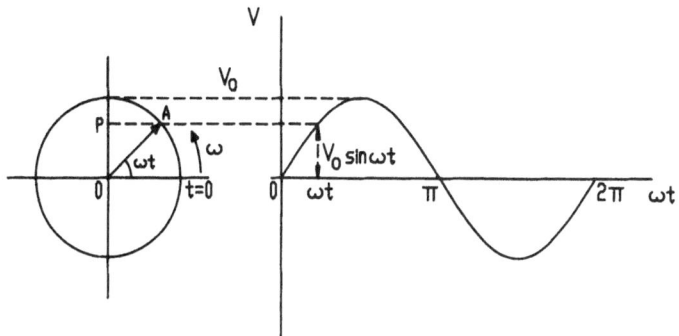

Fig. 4.2 Sine wave generated by the projection of a rotating phasor OA. (Conventionally, anti-clockwise rotation is assumed.)

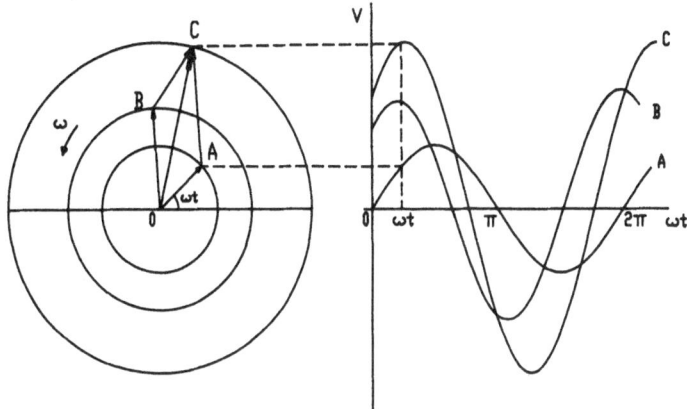

Fig. 4.3 Summation of phasors with $OA \equiv V$, $OB \equiv V'$ and $OC \equiv V''$ such that $\mathbf{OC} = \mathbf{OA} + \mathbf{OB}$. In the curves which are generated, curve A represents $V = V_0 \sin \omega t$, curve B, $V' = V_0' \sin (\omega t + \phi')$, and curve C, $V'' = V_0'' \sin (\omega t + \phi'')$.

the phasor diagram taken at an instant in time, i.e. the *stationary phasor diagram*. The simplest such diagram is shown in Fig. 4.4, from which the resultant phasors may be derived geometrically in terms of

$$V_0'' = (V_0^2 + V_0'^2 + 2V_0 V_0' \cos \phi')^{1/2}$$

and

$$\phi'' = \tan^{-1} \left(\frac{V_0' \sin \phi'}{V_0 + V_0' \cos \phi'} \right)$$

corresponding to (4.7) and (4.8).

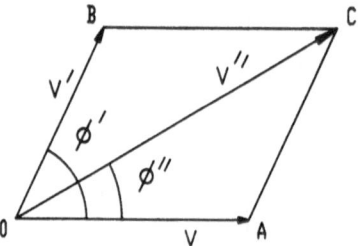

Fig. 4.4 Stationary phasor diagram.

The phasor method can be readily extended to the summation of several voltages or currents. Consider the voltages

$$V_i = V_{i0} \sin(\omega t + \phi_i); \quad i = 1, 2, 3, 4, \ldots$$

each with the same frequency but different amplitudes and phase. If the horizontal axis is taken to correspond to $\phi = 0$ (Fig. 4.5) then the horizontal component of the resultant will be given by

$$V_{HO} = \sum_i V_{i0} \cos \phi_i$$

where the summation includes all of the terms. The vertical component corresponding to $\phi = \pi/2$ will be

$$V_{VO} = \sum_i V_{i0} \sin \phi_i$$

The amplitude of the resultant will be

$$V_{RO} = (V_{HO}^2 + V_{VO}^2)^{1/2} \tag{4.9}$$

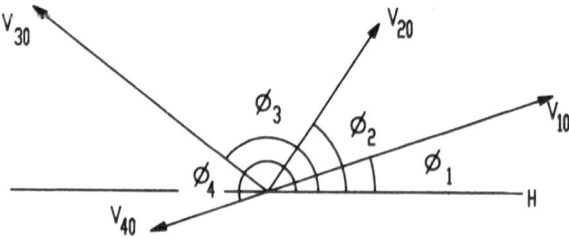

Fig. 4.5 Representation of several phasors.

while its phase angle will be

$$\phi_R = \tan^{-1}\left(\frac{V_{VO}}{V_{HO}}\right) \tag{4.10}$$

Inspection will show that (4.6)–(4.8) are a special case of this result.

4.5 RESISTANCE, SELF-INDUCTANCE AND CAPACITANCE

Consider a sinusoidally varying current $I = I_0 \sin \omega t$ which flows in turn through a resistor, an inductor and a capacitor as shown in Fig. 4.6. The instantaneous voltages required to cause the current to flow through each element are given by the following.

1. For the *resistor* (resistance R)

$$V_R = IR = I_0 R \sin \omega t = V_{RO} \sin \omega t \tag{4.11}$$

2. For the *inductor* (self-inductance L)

$$V_L = L\frac{dI}{dt}$$

$$= L\frac{d}{dt}(I_0 \sin \omega t)$$

$$= I_0 \omega L \cos \omega t = I_0 \omega L \sin\left(\omega t + \frac{\pi}{2}\right)$$

$$= V_{LO} \sin\left(\omega t + \frac{\pi}{2}\right) \tag{4.12}$$

3. For the *capacitor* (capacitance C)

$$V_C = Q/C$$

$$= \frac{\int I\, dt}{C} = \frac{\int I_0 \sin \omega t\, dt}{C} = -\frac{I_0 \cos \omega t}{\omega C} = \frac{I_0}{\omega C} \sin\left(\omega t - \frac{\pi}{2}\right)$$

$$= V_{CO} \sin\left(\omega t - \frac{\pi}{2}\right) \tag{4.13}$$

From (4.11) it is seen that V_R and I are in phase and are simply related by Ohm's law, $V_R = IR$. Equation (4.12) shows that the

Alternating current theory

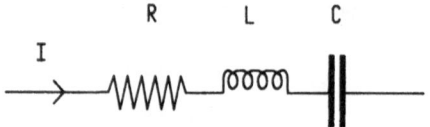

Fig. 4.6 Series R, C, L circuit with $I = I_0 \sin \omega t$.

voltage across the inductor leads the current by $\pi/2$ or $90°$ and that peak voltage is given by

$$V_{L0} = I_0 \omega L = I_0 X_L \qquad (4.14)$$

where $X_L = \omega L$ is called the inductive reactance. Equation (4.13) shows that the voltage V_C across the capacitor lags the current by $\pi/2$ or $90°$ and the peak voltage is given by

$$V_{C0} = \frac{I_0}{\omega C} = I_0 X_C \qquad (4.15)$$

where $X_C = 1/\omega C$ is called the capacitive reactance.

The various voltages for the series circuit can be represented on a voltage phasor diagram (Fig. 4.7(a)), in which the peak current is represented by the horizontal phasor I_0. The peak voltage across the resistor is in phase with the current and the phasor which represents it will therefore be parallel to I_0 and have magnitude $V_{R0} = I_0 R$. However, the peak voltage across the inductor leads the current by $90°$ and the phasor which represents it will be $90°$ 'ahead' of I_0 in the phasor diagram and will have a magnitude of $V_{L0} = I_0 \omega L$. Similarly, the phasor V_{C0} of magnitude $I_0/\omega C$ which is in the negative sense relative to V_{L0} will completely represent the peak voltage across the capacitor, since V_{C0} lags I_0 by $90°$. Figure 4.7(b) illustrates how

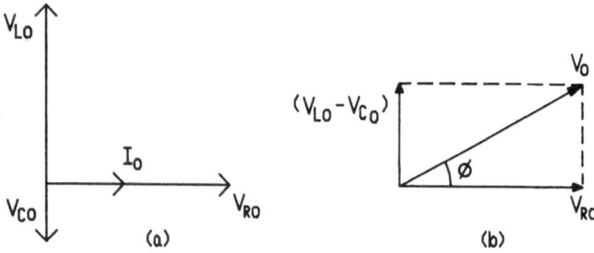

Fig. 4.7 Phasor diagram for R, C, L circuit.

the resultant peak voltage across the series combination of R, L and C can be determined by combining the total vector normal to I_0 namely $V_{L0} - V_{C0}$ with the phasor parallel to I_0 namely V_{R0}: it is seen that the resultant V_0 leads the current by the angle ϕ. Geometrically,

$$V_0 = [V_{R0}^2 + (V_{L0} - V_{C0})^2]^{1/2} = I_0[R^2 + (\omega L - 1/\omega C)^2]^{1/2} \quad \text{(4.16(a))}$$

and

$$\phi = \tan^{-1}\left(\frac{\omega L - 1/\omega C}{R}\right) \quad \text{(4.16(b))}$$

whence the voltage V across the series combination is

$$V = V_0 \sin(\omega t + \phi)$$
$$= I_0|Z|\sin(\omega t + \phi) \quad \text{(4.17)}$$

where

$$|Z| = \left[R^2 + \left(\omega L - \frac{1}{\omega C}\right)^2\right]^{1/2} \quad \text{(4.18)}$$

$|Z|$ is called the *magnitude of the impedance* of the series circuit and represents the ratio of the peak voltage to the peak current, i.e. $|Z| = V_0/I_0$. The reason for writing $|Z|$ will become apparent in section 4.6 below.

Remembering that $X_L = \omega L$ and $X_C = 1/\omega C$, (4.18) can be written as

$$|Z| = (R^2 + X^2)^{1/2} \quad \text{(4.19)}$$

while (4.16(b)) becomes

$$\phi = \tan^{-1}\left(\frac{X}{R}\right) \quad \text{(4.20)}$$

Equations (4.19) and (4.20) are general expressions for the impedance and phase angle of an ac circuit and in both equations X represents the total reactance of the circuit, i.e. $X = X_L - X_C$.

More complicated circuits can also be treated by phasor methods. As a general rule, however, the phasor diagram is best set up relative to that quantity—current or voltage—which is common to the elements of the circuit, i.e. the current for a series circuit, the voltage for a parallel circuit: in the latter case a *current* phasor diagram will result.

4.6 COMPLEX REPRESENTATION OF AC VOLTAGES AND CURRENTS—THE j NOTATION

Treatment of ac circuits is greatly simplified by the use of complex numbers. Use is made of the two well-known results

$$e^{j\theta} = \cos\theta + j\sin\theta$$

and

$$e^{\pm j\pi/2} = \cos\left(\pm\frac{\pi}{2}\right) + j\sin\left(\pm\frac{\pi}{2}\right) = \pm j$$

so that the operator $j = \sqrt{-1}$ is equivalent to a positive or anti-clockwise rotation of $90°$ and $-j$ represents a negative or clockwise rotation of $90°$. The usual notation for $\sqrt{-1}$, namely i, was replaced in network theory by j, to avoid confusion with currents.

The function $\sin\omega t$ can be replaced by the imaginary part of $e^{j\omega t}$ and $\cos\omega t$ by the real part of $e^{j\omega t}$. The mathematical treatment then proceeds as if $e^{j\omega t}$ were the actual time variation.

The method is best illustrated by applying it to the series R, L, C circuit treated in section 4.5. Let the current in Fig. 4.6 be written as

$$I = I_0 e^{j\omega t} = I_0(\cos\omega t + j\sin\omega t) \tag{4.21}$$

The instantaneous total voltage will be

$$V = V_R + V_L + V_C = IR + \frac{L\,dI}{dt} + \frac{\int I\,dt}{C} \tag{4.22}$$

where $\int I\,dt = Q$, the charge. Substituting for I from (4.21) gives

$$V = I_0 e^{j\omega t} R + LI_0 j\omega e^{j\omega t} + \frac{I_0 e^{j\omega t}}{j\omega C}$$

$$= \left(R + j\omega L + \frac{1}{j\omega C}\right)I$$

$$= ZI$$

where Z is the *complex impedance* given by

$$Z = R + j\left(\omega L - \frac{1}{\omega C}\right) \tag{4.23}$$

and V and I are the complex voltage and current respectively.

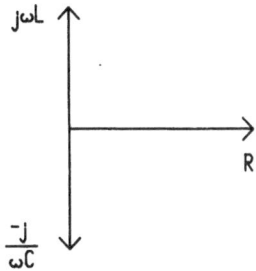

Fig. 4.8 Argand diagram showing the 'rotation' of the impedances due to L and C.

In complex notation, the impedance due to the inductance is $j\omega L$ and that due to the capacitance is $1/j\omega C$ or $-j/\omega C$. On an Argand diagram, in which imaginary quantities are represented by coordinates at right angles to the real axis (Fig. 4.8), the inductive impedance is shown to lead R by $90°$ and the capacitive impedance to lag by the same amount. The lead or lag of $90°$ is represented in the corresponding impedance equation by the positive and negative sign of j, respectively.

The total complex impedance given by (4.23) can be written as

$$Z = |Z| e^{j\phi}$$

whence

$$|Z| = \left[R^2 + \left(\omega L - \frac{1}{\omega L} \right)^2 \right]^{1/2} \quad \text{and} \quad \phi = \tan^{-1} \left(\frac{\omega L - 1/\omega C}{R} \right)$$

which correspond to the values obtained in (4.18) and (4.16(b)) by phasor methods.

In a general circuit the impedance can be written

$$Z = R + jX \tag{4.24}$$

so that $|Z| = (R^2 + X^2)^{1/2}$ and $\phi = \tan^{-1}(X/R)$ as in (4.19) and (4.20), and X is the reactance of the circuit. Thus the time dependence of the voltage V is given by

$$V = ZI = |Z| e^{j\phi} I_0 e^{j\omega t} = |Z| I_0 e^{j(\omega t + \phi)} \tag{4.25}$$

and the phase angle ϕ is derived from the impedance Z. If the actual current flowing is $I_0 \sin \omega t = $ imaginary part$(I_0 e^{j\omega t}) = \text{Im}(I_0 e^{j\omega t})$ then V is the imaginary part of (4.25) and

$$V = |Z| I_0 \sin (\omega t + \phi) \tag{4.26}$$

and $V_0 = I_0|Z|$ as in (4.17). In (4.26), V represents the instantaneous voltage across the circuit. Similarly, if $I = I_0 \cos \omega t = \text{real part}(I_0 e^{j\omega t}) = \text{Re}(I_0 e^{j\omega t})$ then (4.25) gives

$$V = |Z|I_0 \cos(\omega t + \phi) \tag{4.27}$$

with the same values for $|Z|$ and ϕ as above.

4.7 MANIPULATION OF COMPLEX IMPEDANCES

If there are two complex quantities such as $Z_1 = |Z_1|e^{j\phi_1}$ and $Z_2 = |Z_2|e^{j\phi_2}$ then

$$Z_1 Z_2 = (|Z_1||Z_2|)e^{j(\phi_1 + \phi_2)} \tag{4.28}$$

and

$$\frac{Z_1}{Z_2} = \left(\frac{|Z_1|}{|Z_2|}\right)e^{j(\phi_1 - \phi_2)} \tag{4.29}$$

(Thus $|Z_1 Z_2| = |Z_1||Z_2|, |Z_1/Z_2| = |Z_1|/|Z_2|$ since $|e^{j\phi}| = 1$ always.)

For convenience, the complex expression can be abbreviated so that

$$Z = |Z|e^{j\phi} = |Z|/\phi \tag{4.30}$$

Suppose that a voltage $V = V_0 e^{j(\omega t + \phi_1)}$, where V_0 is a real quantity, is applied across an impedance $Z = |Z|e^{j\phi_2}$. The current through the impedance is given by

$$I = \frac{V}{Z} = \frac{V_0 e^{j(\omega t + \phi_1)}}{|Z|e^{j\phi_2}} = \frac{V_0}{|Z|}e^{j(\omega t + \phi_1 - \phi_2)} \tag{4.31}$$

or

$$I = I_0 e^{j(\omega t + \phi)} \tag{4.32}$$

with $I_0 = V_0/|Z|$, a real quantity, and $\phi = \phi_1 - \phi_2$.

Similarly, if the current through a circuit of impedance $|Z|e^{j\phi_2}$ is $I_0 e^{j(\omega t + \phi_1)}$, then the voltage across the impedance is

$$V = I_0|Z|e^{j(\omega t + \phi_1 + \phi_2)} = V_0 e^{j(\omega t + \phi)} \tag{4.33}$$

Since the time variation $e^{j\omega t}$ is common to all terms, the current and voltage are often abbreviated into the form of (4.30) so that

$$V = V_0 e^{j\phi_1} = V_0/\phi_1$$

and

$$Z = |Z|e^{j\phi_2} = |Z|/\phi_2$$

and

$$I = \frac{V}{Z} = \frac{V_0 \underline{/\phi_1}}{|Z| \underline{/\phi_2}} = \frac{V_0}{|Z|} \underline{/\phi_1 - \phi_2} \tag{4.34}$$

4.8 MUTUAL INDUCTANCE IN AN AC CIRCUIT

When the magnetic flux from a coil links with a second coil, a change of current in the first, or primary, coil will cause an emf to be induced in the secondary coil; the value of the induced emf will be determined by the laws of electromagnetic induction. For the present purposes, it is possible to consider two coils, $1°$ and $2°$ in Fig. 4.9, as forming a mutual inductance M. The emf generated in the secondary $2°$ is given by

$$V_2 = - M \frac{dI_1}{dt} \tag{4.35}$$

where I_1 is the current in the primary $1°$. If $I_1 = I_{10}e^{j\omega t}$ then

$$V_2 = - j\omega M I_{10} e^{j\omega t} = - j\omega M I_1 \tag{4.36}$$

Since the coupling of the magnetic field between the two coils is reciprocal, the emf V_1 generated in the primary by an alternating current in the secondary will be given by $V_1 = - j\omega M I_2$.

Clearly the action of the mutual inductance can be represented by an operator $-j\omega M$, where $-j$ indicates a voltage lag of $90°$ behind the current I_1. The peak secondary emf is $\omega M I_{10} = V_{20}$.

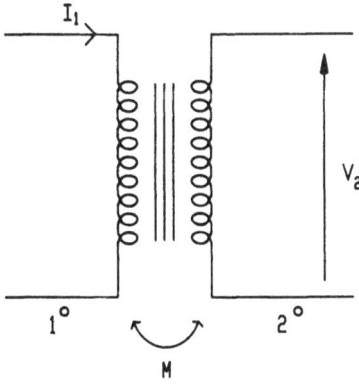

Fig. 4.9 Schematic diagram of a mutual inductance.

4.9 KIRCHHOFF'S LAWS IN AC CIRCUITS

In ac circuits it is necessary to express voltages and currents in terms of their magnitude and phase difference and this can be achieved by expressing them as complex quantities. Thus, in order that the use of Kirchhoff's laws may be extended to ac circuits, it follows that they need to be stated in terms of complex quantities:

1. *current law*—the algebraic sum of the complex currents at any node in a circuit is zero;
2. *voltage law*—in any loop the algebraic sum of the complex potential differences across the impedances in the loop is equal to the algebraic sum of the complex emfs acting in that loop.

4.10 COMPLEX IMPEDANCES IN SERIES AND PARALLEL

The derivation of the total impedance due to a combination of complex impedances is entirely analogous to the dc case treated in section 2.3.

4.10.1 Series circuit

Suppose that a current I flows through three impedances Z_1, Z_2 and Z_3 as shown in Fig. 4.10. The total voltage V will be the sum of the individual voltages across each impedance so that, by Kirchhoff's laws,

$$V = IZ_1 + IZ_2 + IZ_3 \qquad (4.37)$$

If the combination of impedances is to be equivalent to a single impedance Z then

$$V = IZ \qquad (4.38)$$

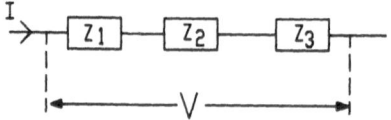

Fig. 4.10 Current I flowing in turn through Z_1, Z_2 and Z_3.

Thus from (4.37) and (4.38)

$$Z = Z_1 + Z_2 + Z_3 \qquad (4.39)$$

The general case will be

$$Z = \sum_n Z_n. \qquad (4.40)$$

4.10.2 In parallel

If three impedances are connected in parallel (Fig. 4.11), a voltage V placed across them will generate currents I_1, I_2 and I_3 which, by Kirchhoff's laws, must in sum equal the total current I flowing into the circuit and must also satisfy the condition that $V = I_1 Z_1 = I_2 Z_2 = I_3 Z_3$. Thus

$$I = I_1 + I_2 + I_3 = \frac{V}{Z_1} + \frac{V}{Z_2} + \frac{V}{Z_3} \qquad (4.41)$$

If the parallel combination of impedances is to be equivalent to a single impedance Z then

$$I = V/Z \qquad (4.42)$$

and, from (4.41) and (4.42),

$$\frac{1}{Z} = \frac{1}{Z_1} + \frac{1}{Z_2} + \frac{1}{Z_3} \qquad (4.43)$$

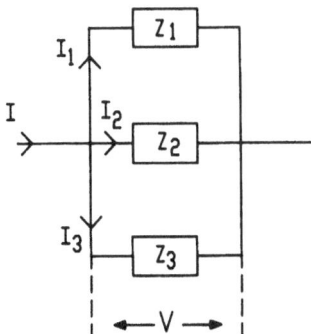

Fig. 4.11 Current divided between three parallel impedances.

The general case will, therefore, be

$$\frac{1}{Z} = \sum_n \frac{1}{Z_n} \tag{4.44}$$

4.11 ADMITTANCE OF AN AC CIRCUIT

The reciprocal of the impedance Z of a circuit is defined as the *admittance* Y of the circuit; thus

$$Y = 1/Z \tag{4.45}$$

and, by definition, the current and voltage are related by

$$I = YV \tag{4.46}$$

Since $Z = R + jX$

$$Y = \frac{1}{R + jX} = \frac{R - jX}{R^2 + X^2} \tag{4.47}$$

The admittance Y can be expressed in terms of a *conductance* G and a *susceptance* B so that

$$Y = G + jB \tag{4.48}$$

Hence, from (4.47) and (4.48)

$$G = \frac{R}{R^2 + X^2} \quad \text{and} \quad B = \frac{-X}{R^2 + X^2} \tag{4.49}$$

(Note that only if $X = 0$ will $G = 1/R$ and only if $R = 0$ will $B = -1/X$.)

It is generally more convenient to work in terms of admittances when a number of elements are connected in parallel and also when applying nodal analysis to a circuit.

4.12 ROOT MEAN SQUARE VALUES OF AC QUANTITIES

If, at a given instance, a current I flows in a resistor of resistance R, the instantaneous rate of energy loss due to Joule heating is

$$p = I^2 R$$

If the current varies over a period of time, the *mean power*, $\bar{p}$, dissipated can be expressed as

$$\bar{p} = \overline{I^2 R} = \overline{I^2} R$$

where the bar indicates the mean value over the given period of time and $\overline{I^2}$ is called the mean square value of I.

For the purpose of considering power dissipation, an effective current may be defined as I_{rms}, the *root mean square* or *rms* current, where

$$I_{rms} = \sqrt{\overline{I^2}} \tag{4.50}$$

Thus

$$\bar{p} = I_{rms}^2 R \tag{4.51}$$

I_{rms} is the magnitude of the dc current which would produce the same average Joule heating as the varying current I. It is usual, therefore, to quote ac voltages and currents in terms of their rms values; so that when the emf of the mains supply is quoted as 240 V, it is understood that this is actually the rms value.

For a current $I = I_0 \sin \omega t$

$$I_{rms}^2 = \overline{I^2} = \overline{I_0^2 \sin^2 \omega t} = \tfrac{1}{2} I_0^2 \overline{(1 - \cos 2\omega t)}$$

Over a complete cycle, $\cos 2\omega t$ averages to zero so that

$$I_{rms}^2 = \tfrac{1}{2} I_0^2$$

or

$$I_{rms} = I_0/\sqrt{2} \quad \text{and} \quad I_0 = \sqrt{2} I_{rms} \tag{4.52}$$

Similarly, it follows that

$$V_{rms} = V_0/\sqrt{2} \quad \text{and} \quad V_0 = \sqrt{2} V_{rms} \tag{4.53}$$

Thus a voltage of 240 V (rms) will have a peak value of about 340 V.

From (4.52) and (4.53), it follows that any relationship established between I_0 and V_0 will equally apply between I_{rms} and V_{rms}; for example, if $V_0 = I_0|Z|$ then $V_{rms} = I_{rms}|Z|$.

4.13 POWER IN AC CIRCUITS

The current I which flows through an impedance, or a network of impedances, will generally not be in phase with the voltage V across it. This fact can be expressed by writing $V = V_0 \sin \omega t$ and $I = I_0 \sin(\omega t - \phi)$ where $-\pi/2 \leqslant \phi \leqslant \pi/2$.

The instantaneous power generated in the impedance will

then be

$$p = VI = V_0 I_0 \sin \omega t \sin (\omega t - \phi)$$
$$= V_0 I_0 (\sin^2 \omega t \cos \phi - \sin \omega t \cos \omega t \sin \phi)$$
$$= \tfrac{1}{2} V_0 I_0 [(1 - \cos 2\omega t) \cos \phi - \sin 2\omega t \sin \phi] \qquad (4.54)$$

The component $\tfrac{1}{2} V_0 I_0 (1 - \cos 2\omega t) \cos \phi$ is never less than zero and is called the *instantaneous active power*: it represents the power transferred from the generator to the impedance or network at the given time t. The component $-\tfrac{1}{2} V_0 I_0 \sin 2\omega t \sin \phi$ can have either sign and is called the *instantaneous reactive power* and represents the continual interchange of power between the generator and the reactive part of the impedance or network; the time average of this term is zero.

The *average power* supplied to the network or impedance over one cycle will be

$$P = \tfrac{1}{2} V_0 I_0 \cos \phi = V_{rms} I_{rms} \cos \phi \qquad (4.55)$$

since

$$\overline{(1 - \cos 2\omega t)} = 1$$

The product $V_{rms} I_{rms}$ is called the *apparent power S*, and is measured in *voltamperes* (VA) to distinguish it from real power which is measured in watts. The quantity $\cos \phi$ is known as the *power factor* of the impedance or network and $0 \leqslant \cos \phi \leqslant 1$ giving $\cos \phi = 1$ for a pure resistance, $\cos \phi = 0$ for a pure reactance.

For completeness, since the power factor does not indicate the sign of ϕ, it is usual to state the power factor as either a lagging ($\phi > 0$) or a leading ($\phi < 0$) power factor.

The *reactive power Q*, which is said to be in quadrature with the real or average power, is given by

$$Q = \tfrac{1}{2} V_0 I_0 \sin \phi = V_{rms} I_{rms} \sin \phi \qquad (4.56)$$

and is measured in *var* (voltamperes reactive) or, in practical use, kvar. Since $\sin \phi$ can be positive or negative, it is usual to quote Q as a positive quantity and indicate its value in var (inductive) for $\phi > 0$ and var (capacitive) for $\phi < 0$.

To summarize, the apparent power is

$$S = V_{rms} I_{rms} = I_{rms}^2 |Z| \, \text{VA}$$

the average power is

$$P = V_{rms} I_{rms} \cos \phi = I_{rms}^2 R \, \text{W}$$

the reactive power is

$$Q = V_{rms}I_{rms} \sin \phi = I_{rms}^2 X \text{ var}$$

and

$$S^2 = P^2 + Q^2 \tag{4.57}$$

From (4.54) the power at an instant t can be written

$$p = P(1 - \cos 2\omega t) - Q \sin 2\omega t \tag{4.58}$$

and the frequency of p is twice that of V or I.

Figure 4.12(a) represents the variation of the current and voltage in a circuit such that the current lags the voltage by angle ϕ. The corresponding instantaneous power p is shown in Fig. 4.12(b) and it should be noted that, when p is positive, energy is being transferred from the generator to the circuit and, when p is negative, energy is being returned from the circuit to the generator.

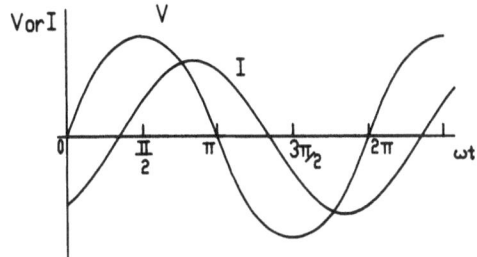

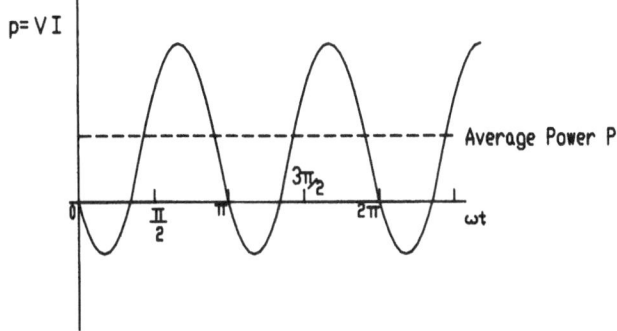

Fig. 4.12 Instantaneous power p for voltage $V = V_0 \sin \omega t$ and current $I = I_0 \sin (\omega t - \phi)$.

4.14 COMPLEX POWER AND THE POWER TRIANGLE

In order to express the power in a circuit in complex form, the dc equivalent voltage is written as

$$V = V_{rms}\, e^{j\phi_1}$$

and the current as

$$I = I_{rms}\, e^{j\phi_2}$$

where both V_{rms} and I_{rms} are taken to be real.

Because it is the difference in phase angle which determines the power factor, the *complex power* S' is obtained by using the complex conjugate of I, i.e.

$$S' = VI^* = V_{rms}e^{j\phi_1}I_{rms}e^{-j\phi_2} = V_{rms}I_{rms}\, e^{j\phi} \qquad (4.59)$$

where ϕ is the phase difference. From (4.59), the complex power S' becomes

$$S' = V_{rms}I_{rms}\cos\phi + jV_{rms}I_{rms}\sin\phi = P + jQ \qquad (4.60)$$

where P and Q are, from (4.55) and (4.56), the average and reactive power respectively. Since $|S'| = (P^2 + Q^2)^{1/2} = S$, the complex power may be written as

$$S' = S\underline{/\phi} = S\underline{/\phi_1 - \phi_2} \qquad (4.61)$$

The components of the power may thus be represented diagrammatically as in Fig. 4.13, where the figures ABC are said to be the *power triangles*.

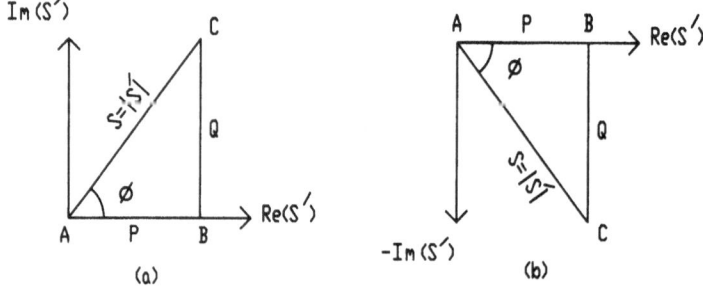

Fig. 4.13 Representation of the complex power: (a) $\cos\phi$ lagging, Q inductive; (b) $\cos\phi$ leading, Q capacitive.

4.15 POWER USAGE: IMPROVEMENT OF THE POWER FACTOR

Consider a supply voltage V_{rms} connected to a number of impedances connected in parallel as in Fig. 4.14. Each impedance can be regarded as representing, for example, the electrical network of a private or commercial consumer.

The total complex power S_T supplied by the source is

$$S'_T = V_{rms}I_{1rms} + V_{rms}I_{2rms} + \cdots + V_{rms}I_{nrms}$$

which, from (4.59), becomes

$$S'_T = P_1 + jQ_1 + P_2 + jQ_2 + \cdots + P_n + jQ_n = P_T + jQ_T \quad (4.62)$$

where P_T is the total average power and Q_T is the total reactive power.
The total apparent power $S_T = |S'_T|$ is given by

$$S_T = (P_T^2 + Q_T^2)^{1/2} \quad (4.63)$$

while the total power factor is

$$\cos \phi_T = P_T/Q_T \quad (4.64)$$

(It is a simple matter to show that (4.62), (4.63) and (4.64) apply equally to impedances connected in series.)

Equation (4.62) shows that the power triangles for the individual impedances or networks can be connected in sequence, as shown in Fig. 4.15, to form a total power triangle, ABC. The figure has been drawn for Q_1 and Q_2 inductive, Q_4 capacitive, whilst Q_3 is zero indicating a purely resistive load. The total power triangle has $AB = P_T$, $BC = Q_T$ and $AC = S_T$ whilst $\underline{/BAC} = \phi_T$ corresponds to a total power factor $\cos \phi_T$ lagging.

The suppliers of mains electricity rate their alternators, transformers and power lines in terms of apparent power, kVA or MVA,

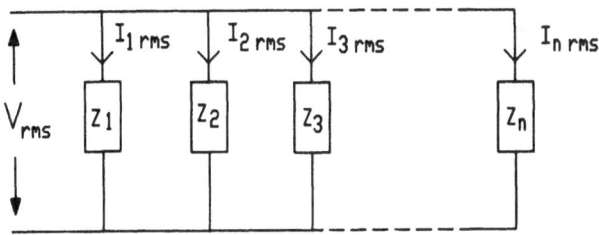

Fig. 4.14 Power supplied to networks in parallel.

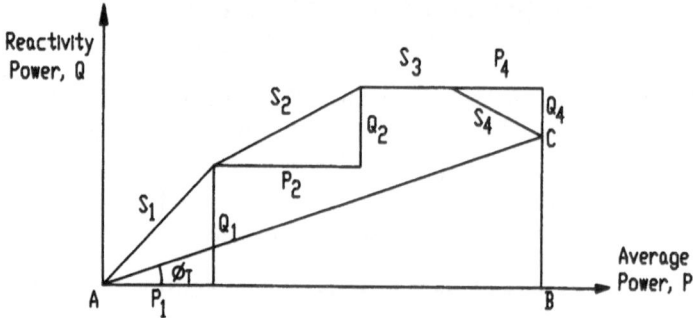

Fig. 4.15 The total power triangle.

rather than average power (kW or MW) and charge accordingly. Any consumer, therefore, who operates equipment which has a complex impedance as opposed to a pure resistance, is charged for more than the real energy consumed. Thus, from the point of view of both the supplier and the consumer, energy is wasted if the power factor differs from unity.

Much electrical equipment has a reactive power component and, for most, this component is inductive. For large industrial users, there may be advantages in 'improving the power factor' by seeking to cancel out the reactive power by introducing a network with a Q of the opposite sign: in most cases this would mean introducing a bank of capacitors in parallel with the source of supply to reduce the inductive component. If Q_T could be made zero, the running costs to the consumer would be reduced, thus enabling some of the generating capabilities of the supplier to be made available to other customers. However, the installation of such a capacitor bank rests solely on the economic decision of the consumer, since he has to compare the cost of installation with the savings to be realized.

4.16 IMPEDANCE MATCHING

When a voltage generator which is connected to a load of impedance Z_L has an internal impedance Z_G, the power delivered to the load will depend on the relationship of Z_G to Z_L. When the power delivered is a maximum, the load is said to be *matched*.

Consider the simplest case shown as in Fig. 4.16 and let

$$Z_G = R_G + jX_G \quad \text{and} \quad Z_L = R_L + jX_L$$

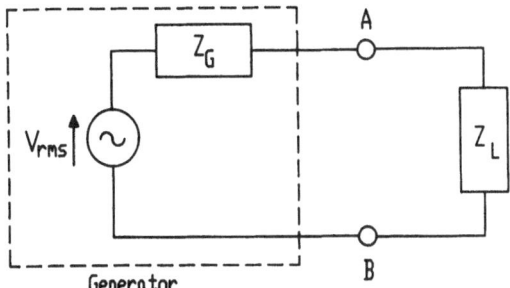

Fig. 4.16 Generator with internal impedance Z_G connected to a load Z_L.

The rms current will be

$$I_{rms} = \frac{V_{rms}}{|Z_G + Z_L|} = V_{rms}[(R_G + R_L)^2 + (X_G + X_L)^2]^{-1/2}$$

The average power dissipated in the load will be

$$P = I_{rms}^2 R_L = V_{rms}^2 R_L[(R_G + R_L)^2 + (X_G + X_L)^2]^{-1} \quad (4.65)$$

It is possible to think of maximizing the power under three practical conditions: either X_L is fixed and R_L is variable, or R_L and X_L are both variable but their ratio is fixed, i.e. the phase angle ϕ_L of the load is fixed, or R_L and X_L are both independently variable.

1. X_L is fixed and R_L is variable.
   From (4.65)

$$\frac{dP}{dR_L} = \frac{V_{rms}^2[(R_G + R_L)^2 + (X_G + X_L)^2] - V_{rms}^2 R_L \times 2(R_G + R_L)}{[(R_G + R_L)^2 + (X_G + X_L)^2]^2}$$

For maximum transfer of power to the load $dP/dR_L = 0$. Therefore

$$(R_G + R_L)^2 + (X_G + X_L)^2 - 2R_L(R_G + R_L) = 0$$

which is satisfied if

$$R_L = [R_G^2 + (X_G + X_L)^2]^{1/2} \quad (4.66)$$

From (4.66), if $X_L = 0$ then

$$R_L = |Z_G| \quad (4.67)$$

2. R_L and X_L are both variable but their ratio X_L/R_L is a constant; $k = \tan \phi_L$, where ϕ_L is the phase angle of the load.

Since $X_L = kR_L$, (4.65) becomes

$$P = V_{rms}^2 R_L[(R_G + R_L)^2 + (X_G + kR_L)^2]^{-1}$$

For maximum power $dP/dR_L = 0$ which requires that

$$(R_G + R_L)^2 + (X_G + kR_L)^2 - R_L[2(R_G + R_L) + 2k(X_G + kR_L)] = 0$$

and this condition is satisfied if

$$R_L = \left(\frac{R_G^2 + X_G^2}{1 + k^2}\right)^{1/2} = |Z_G| \cos \phi_L \qquad (4.68)$$

3. R_L and X_L are independently variable.

It is obvious from (4.65) that, in terms of X_L, the condition for maximum P is $X_L = -X_G$. This may be verified by putting $(\partial P/\partial X_L)_{R_L} = 0$.

The expression for the power reduces to an equivalent of the dc expression, (2.14).

$$P = \frac{V_{rms}^2 R_L}{(R_G + R_L)^2}$$

which gives a maximum when $R_L = R_G$.

It is clear that the maximum power transferable from the generator under these conditions occurs when the load impedance satisfies the conditions $R_L = R_G$ and $X_L = X_G$ or

$$Z_L = Z_G^* \qquad (4.69)$$

i.e. the impedance of the load is equal to the complex conjugate of the impedance of the source.

4.17 COMPLEX FREQUENCY—s NOTATION

The analysis of electrical circuits can usefully be extended to the case where the voltage or current variation $e^{j\omega t}$ is replaced by e^{st} where

$$s = j\omega + \sigma = j(\omega - j\sigma)$$

which corresponds to a *complex frequency*, $\omega - j\sigma$.

σ is the real part of s and is called the *neper frequency*: it is measured in *nepers per second* and corresponds to the exponential growth or decay in a signal such that, when $\sigma t = 1$ neper, the signal amplitude changes by a factor e. (Normally, $\sigma < 0$, corresponding to an exponential decay.)

Consider the voltage

$$V = V_{co} e^{st} \qquad (4.70)$$

where V_{co} is the complex amplitude of the voltage such that $V_{co} = V_0 e^{j\phi}$ and ϕ corresponds to a phase difference. Then

$$V = V_0 e^{j\phi} e^{(j\omega + \sigma)t} = V_0 e^{\sigma t} e^{j(\omega t + \phi)}$$
$$= V_0 e^{\sigma t} [\cos(\omega t + \phi) + j \sin(\omega t + \phi)] \qquad (4.71)$$

Thus

$$\mathrm{Re}(V) = V_0 e^{\sigma t} \cos(\omega t + \phi) \qquad (4.72)$$

and

$$\mathrm{Im}(V) = V_0 e^{\sigma t} \sin(\omega t + \phi) \qquad (4.73)$$

If $\phi = 0$ (4.70) becomes

$$V = V_0 e^{st} \qquad (4.74)$$

Hence

for $s = 0, V = V_0$ i.e. a dc voltage

for $s = \sigma, V = V_0 e^{\sigma t}$ i.e. an exponential voltage (usually $\sigma < 0$ corresponding to a decay)

for $s = j\omega, V = V_0 e^{j\omega t}$ i.e. a sinusoidal voltage

for $s = j\omega + \sigma, V = V_0 e^{(j\omega + \sigma)t}$ i.e. an exponentially varying sinusoidal voltage.

Obviously it is also possible to express current in the same form, and the various ac equations may be generalized as follows:

1. $V = V_{co} e^{st}$ and $I = I_{co} e^{st}$ (4.75)

2. $V = IR$ and $I = GV$ (4.76)

3. $V = L\dfrac{dI}{dt} = L\dfrac{d}{dt}(I_{co} e^{st}) = sLI$ and $I = \dfrac{V}{sL}$ (4.77)

4. $V = \dfrac{Q}{C} = \dfrac{\int I\,dt}{C} = \dfrac{\int I_{co} e^{st}\,dt}{C} = \dfrac{I}{sC}$ and $I = sCV$ (4.78)

5. $Z(s) = \dfrac{V(s)}{I(s)}$ and $Y(s) = \dfrac{I(s)}{V(s)}$ (4.79)

where $Z(s)$ and $Y(s)$ are the generalized impedance and admittance respectively.

5

Mesh or loop analysis and nodal analysis

The application of Kirchhoff's laws to any circuit which consists of more than one mesh is best achieved by either mesh (loop) analysis or nodal analysis. In this chapter, the reader is introduced to both types of analysis by first considering their application to a simple circuit: in both instances, the analysis is extended to a generalized circuit in which a generalized notation is used and consequently a generalized system of solution is achieved. Finally, mesh analysis is extended to circuits in which different loops are coupled together by mutual inductances.

5.1 MESH (LOOP) ANALYSIS

Mesh (loop) analysis can be applied to both ac and dc circuits as long as the passive elements can be expressed as impedances or resistances and the sources as voltage generators. Consider the circuit shown in Figs 5.1(a) and 5.1(b). In the former, (a), the two meshes are allocated currents I_1, and I_2, each circulating in the same (clockwise) sense. In the latter, (b), the first mesh is allocated a clockwise circulating current I'_1, whilst the outer loop is allocated a current I'_2, circulating in the anticlockwise sense.

It will be noted that, in the mesh analysis of the circuit shown in Fig. 5.1(a), an impedance such as Z_1 or Z_3, which is associated with only one mesh, carries a current equal to the mesh current; whilst an impedance such as Z_2, which is common to two meshes, carries a current equal to the difference between the currents in the two meshes. Thus, whilst Z_1 carries current I_1, Z_2 carries current $I_1 - I_2$ in the sense of I_1, or $I_2 - I_1$ in the sense of I_2. There are occasions when it is possible to simplify calculations by using 'loop' as well

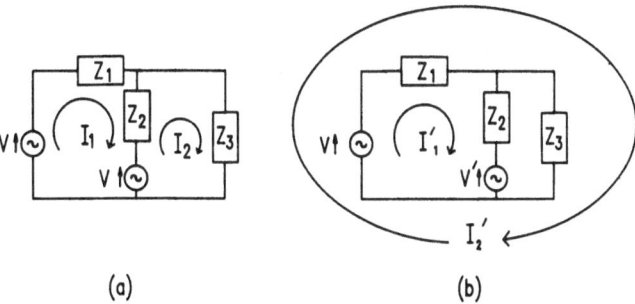

Fig. 5.1 (a) The allocation of mesh currents to a circuit. (b) The allocation of mesh and loop currents to a circuit.

as 'mesh' currents. Thus, if the current in Z_2 is to be calculated, then, by allocating the currents as shown in Fig. 5.1(b), only I_1' needs to be found, whereas in Fig. 5.1(a) both I_1 and I_2 must be determined. In the case of the circuit shown in Fig. 5.1(b), however, an impedance such as Z_1 is common to the first mesh and the loop, and it is allocated a current $I_1' - I_2'$ in the sense of I_1' or $I_2' - I_1'$ in the sense of I_2'. It is clear that the use of circulating currents ensures that Kirchhoff's current law is correctly applied.

Returning to the circuit shown in Fig. 5.1(a), Kirchhoff's voltage law may be applied to each mesh to give a set of linear equations. Thus for the I_1 mesh

$$I_1 Z_1 + (I_1 - I_2)Z_2 = I_1(Z_1 + Z_2) - I_2 Z_2 = V - V' \qquad (5.1)$$

and for the I_2 mesh

$$(I_2 - I_1)Z_2 + I_2 Z_3 = -I_1 Z_2 + I_2(Z_2 + Z_3) = V' \qquad (5.2)$$

Equations (5.1) and (5.2) show that, provided all the mesh currents are in the opposite sense to the loop current, the mesh and loop equations can be formulated more readily by applying the following rule, which sounds more complicated than it is.

> for any mesh (loop), the product of the mesh current and the sum of all the impedances in that mesh minus the sum of the products of the current in any adjacent mesh and the impedance of the branch sharing that current, is equal to the algebraic sum of the emfs acting in the mesh when those emfs acting in the same sense as the mesh current

are considered to be positive and those acting in the opposite sense are considered to be negative.

The application of this rule to most circuits enables the resulting mesh (loop) equations to be expressed directly in matrix form. Thus, for the circuit shown in Fig. 5.1(a)

$$\begin{bmatrix} Z_1 + Z_2 & -Z_2 \\ -Z_2 & Z_2 + Z_3 \end{bmatrix} \begin{bmatrix} I_1 \\ I_2 \end{bmatrix} = \begin{bmatrix} V - V' \\ V' \end{bmatrix} \tag{5.3}$$

whilst the alternative circuit of Fig. 5.1(b) gives

$$\begin{bmatrix} Z_1 + Z_2 & -Z_1 \\ -Z_1 & Z_1 + Z_3 \end{bmatrix} \begin{bmatrix} I_1' \\ I_2' \end{bmatrix} = \begin{bmatrix} V - V' \\ -V \end{bmatrix} \tag{5.4}$$

The solution of these matrix equations is discussed in the next section when a more generalized circuit is considered.

5.2 MESH AND LOOP ANALYSIS APPLIED TO A GENERALIZED CIRCUIT

Equation (5.4) is a particular case of a general equation of the type

$$\begin{bmatrix} Z_{11} & Z_{12} \\ Z_{21} & Z_{22} \end{bmatrix} \begin{bmatrix} I_1 \\ I_2 \end{bmatrix} = \begin{bmatrix} V_1 \\ V_2 \end{bmatrix} \tag{5.5}$$

where Z_{11} and Z_{22} are the total impedances through which only I_1 and I_2 respectively pass. Z_{12} is the total impedance through which both I_1 and I_2 pass and it is positive if I_1 and I_2 have the same sense through the impedance and it is negative if the two currents are of opposing sense. It is obvious that $Z_{12} = Z_{21}$ since Z_{21} is also the total impedance through which both I_1 and I_2 pass. V_1 and V_2 correspond to the algebraic sums of the emfs acting in the meshes carrying the currents I_1 and I_2 respectively and the sign of an individual emf acting in the mesh is taken to be positive if it is in the same sense as the mesh current and negative if it acts in the opposite sense.

Similarly, (5.4) can be generalized as

$$\begin{bmatrix} Z_{11} & Z_{12}' \\ Z_{21}' & Z_{22}' \end{bmatrix} \begin{bmatrix} I_1' \\ I_2' \end{bmatrix} = \begin{bmatrix} V_1 \\ V_2' \end{bmatrix} \tag{5.6}$$

where only the elements which carry the loop current I_2' and the corresponding emf V_2' are changed.

Clearly the above notation can be extended to a generalized circuit containing n meshes (loops) carrying current I_1 to I_n, so that

$$\begin{bmatrix} Z_{11} & Z_{12} & \cdots & Z_{1k} & \cdots & Z_{1n} \\ Z_{21} & Z_{22} & \cdots & Z_{2k} & \cdots & Z_{2n} \\ \vdots & \vdots & & \vdots & & \vdots \\ Z_{j1} & Z_{j2} & \cdots & Z_{jk} & \cdots & Z_{jn} \\ \vdots & \vdots & & \vdots & & \vdots \\ Z_{n1} & Z_{n2} & \cdots & Z_{nk} & \cdots & Z_{nn} \end{bmatrix} \begin{bmatrix} I_1 \\ I_2 \\ \vdots \\ I_j \\ \vdots \\ I_n \end{bmatrix} = \begin{bmatrix} V_1 \\ V_2 \\ \vdots \\ V_j \\ \vdots \\ V_n \end{bmatrix} \tag{5.7}$$

where Z_{jj} is the total impedance through which I_j alone passes and V_j is the algebraic sum of the emfs acting in the jth mesh (loop) where the signs of such emfs are taken to be positive if they act in the same sense as the mesh (loop) current and negative if they act in the opposite sense. $Z_{jk} = Z_{kj}$ which, for $j \neq k$, represents the total impedance of the branch which is common to both the jth and kth meshes (loops) and if the corresponding mesh (loop) currents, I_j and I_k, flow in the same sense through the common branch then Z_{jk} will be positive, but if they flow in opposite sense then Z_{jk} will be negative.

The solutions of (5.7) are given by Cramer's rule (see C. W. Evans, *Engineering Mathematics*, Chapman and Hall, London, 1989, p. 328) so that

$$I_j = \frac{\Delta_j}{\Delta} \tag{5.8}$$

where Δ is the value of the determinant from the impedance matrix in (5.7), i.e.

$$\Delta = \begin{bmatrix} Z_{11} & Z_{12} & \cdots & Z_{1j} & \cdots & Z_{1n} \\ \vdots & \vdots & & \vdots & & \vdots \\ Z_{j1} & Z_{j2} & \cdots & Z_{jj} & \cdots & Z_{jn} \\ \vdots & \vdots & & \vdots & & \vdots \\ Z_{n1} & Z_{n2} & \cdots & Z_{nj} & \cdots & Z_{nn} \end{bmatrix} \tag{5.9}$$

whilst Δ_j is the value of the determinant found by replacing the jth column in the impedance matrix, i.e. Z_{1j} to Z_{nj}, by the net emfs V_1 to V_n, so that

$$\Delta_j = \begin{bmatrix} Z_{11} & Z_{12} & \cdots & V_1 & Z_{1,j+1} & \cdots & Z_{1n} \\ \vdots & \vdots & & \vdots & \vdots & & \vdots \\ Z_{j1} & Z_{j2} & \cdots & V_j & Z_{j,j+1} & \cdots & Z_{jn} \\ \vdots & \vdots & & \vdots & \vdots & & \vdots \\ Z_{n1} & Z_{n2} & \cdots & V_n & Z_{n,j+1} & \cdots & Z_{nn} \end{bmatrix} \tag{5.10}$$

In matrix form $[I] = [Z]^{-1}[V]$ or

$$
\begin{bmatrix} I_1 \\ I_2 \\ \vdots \\ I_j \\ \vdots \\ I_n \end{bmatrix} = \frac{1}{\Delta} \begin{bmatrix} \Delta_{11} & \Delta_{21} & \cdots & \Delta_{k1} & \cdots & \Delta_{n1} \\ \Delta_{12} & \Delta_{22} & \cdots & \Delta_{k2} & \cdots & \Delta_{n2} \\ \vdots & \vdots & & \vdots & & \vdots \\ \Delta_{1j} & \Delta_{2j} & \cdots & \Delta_{kj} & \cdots & \Delta_{nj} \\ \vdots & \vdots & & \vdots & & \vdots \\ \Delta_{1n} & \Delta_{2n} & \cdots & \Delta_{kn} & \cdots & \Delta_{nn} \end{bmatrix} \begin{bmatrix} V_1 \\ V_2 \\ \vdots \\ V_j \\ \vdots \\ V_n \end{bmatrix}
\tag{5.11}
$$

where Δ_{kj} is the cofactor corresponding to the element Z_{kj} in the determinant in (5.9), and is equal to the product of $(-1)^{k+j}$ and the determinant formed by deleting the kth row and jth column from the determinant Δ.

To realize how straightforward this method is, the reader should first show that (5.1) and (5.2) give the results

$$
I_1 = \frac{V(Z_2 + Z_3) - V'Z_3}{Z_1Z_2 + Z_2Z_3 + Z_3Z_1}
$$

$$
I_2 = \frac{VZ_2 + V'Z_1}{Z_1Z_2 + Z_2Z_3 + Z_3Z_1}
$$

both by algebraic elimination and by solving (5.3), and then extend the circuit of Fig. 5.1 to three meshes or two meshes and a loop and set up the corresponding matrices.

5.3 APPLICATION OF MESH (LOOP) ANALYSIS TO CIRCUITS CONTAINING MUTUAL INDUCTANCES

When mesh (loop) analysis is applied to circuits in which there are mutual inductances, the sense of the two windings must be indicated, since the reversal of one current winding will reverse the sense of any mutually induced emf.

In the cases illustrated in Figs 5.2(a) and 5.2(b) the positions of the two dots indicate the relative direction of the windings of the two coils. In both cases the direction of the self-induced voltages are fixed such that they act in the opposite direction to the currents producing them. In Fig. 5.2(a), the dots indicate that the windings of the two coils are in the same sense and, for currents passing in the same direction along the coils, i.e. both currents enter or leave at the ends marked by a dot, the mutually induced emfs act in the same direction as the self-induced emfs. In Fig. 5.2(b), the winding of one

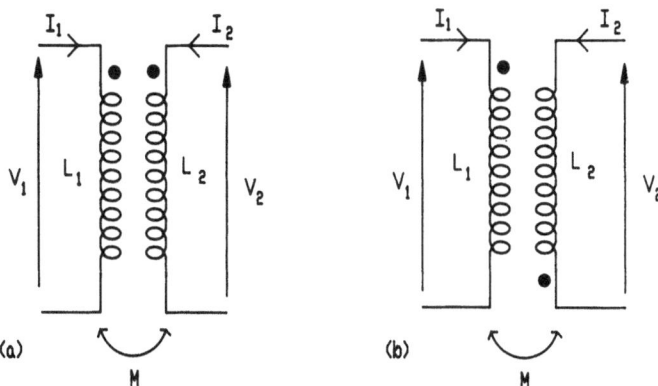

Fig. 5.2 Mutual inductances (a) with dots indicating coils wound in the same sense and (b) with dots indicating coils wound in opposite senses.

coil has been reversed, i.e. the current in one coil enters at an end marked by a dot and the current in the other coil leaves at an end marked by a dot, and mutually induced emfs now act in the opposite direction to the self-induced emfs. Hence, the voltages V_1 and V_2 will be given by

$$V_1 = j\omega L_1 I_1 \pm j\omega M I_2$$
$$V_2 = j\omega L_2 I_2 \pm j\omega M I_1$$

(5.12)

where the positive sign corresponds to the situation illustrated in Fig. 5.2(a) and the negative sign to that shown in Fig. 5.2(b).

By applying (5.12) to coupled meshes, it is possible to take account of the relative directions of the windings of the mutual inductances, but this additional parameter makes it difficult to formulate the mesh (loop) equations directly in matrix form in the way proposed in section 5.1. In such cases it is easier to derive each mesh (loop) equation separately before establishing the corresponding matrix. The method is illustrated with reference to the circuit shown in Fig. 5.3. For the left-hand mesh, the two currents have opposite senses for M_1 and the same sense for M_2 so that

$$(R + j\omega L_1)I_1 + (R_3 + j\omega L_3)(I_1 - I_2) - j\omega M_1 I_2 + j\omega M_2 I_2 = V$$

or

$$[R_1 + R_3 + j\omega(L_1 + L_3)]I_1 - [R_3 + j\omega(L_3 + M_1 - M_2)]I_2 = V$$

(5.13)

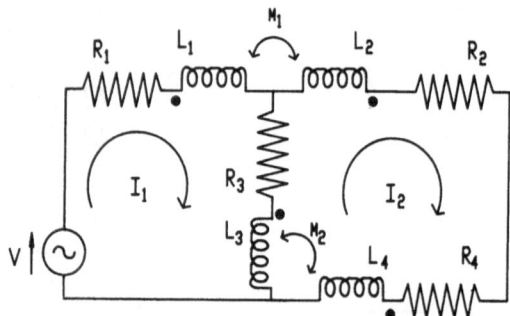

Fig. 5.3 Circuit containing mutual inductances.

For the right-hand mesh, the two currents have opposite senses for both M_1 and M_2, so that

$$(R_2 + j\omega L_2 + R_4 + j\omega L_4)I_2 + (R_3 + j\omega L_3)(I_2 - I_1)$$
$$- j\omega M_1 I_1 - j\omega M_2(I_2 - I_1) - j\omega M_2 I_2 = 0$$

or

$$- [R_3 + j\omega(L_3 + M_1 - M_2)]I_1$$
$$+ [R_2 + R_3 + R_4 + j\omega(L_2 + L_3 + L_4 - 2M_2)]I_2 = 0 \qquad (5.14)$$

Equations (5.13) and (5.14) can be written in matrix form and subsequently solved.

5.4 INPUT IMPEDANCE OF A NETWORK

Consider a voltage source of emf V_s and internal impedance Z_s placed across the terminals AB of the jth mesh of a linear, n-mesh network in which all other voltage sources have been replaced by their internal impedances as indicated in Fig. 5.4(a). If the current from the source is I_j and the pd across the terminals AB is V_j, then the network can be replaced by an equivalent circuit as shown in Fig. 5.4(b).

From Fig. 5.4(b), the *input* or *driving point impedance*, Z_j^{IN}, can be defined by the equation

$$Z_j^{\text{IN}} = V_j/I_j \qquad (5.15)$$

Since in the network all voltage sources have been replaced by their

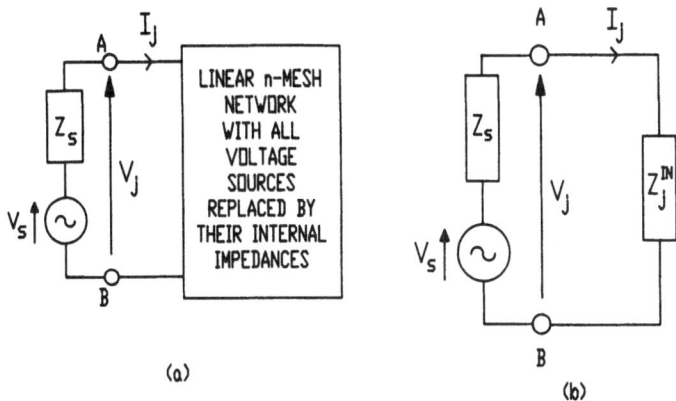

Fig. 5.4 (a) Voltage source placed across the terminals AB of the jth mesh of a linear n-mesh network, and (b) the equivalent circuit.

internal impedances, i.e. $V_k = 0$, except for $k = j$, (5.11) gives

$$I_j = \frac{\Delta_{jj}}{\Delta} V_j$$

which, by comparison with (5.15), leads to

$$Z_j^{IN} = \frac{\Delta}{\Delta_{jj}} \qquad (5.16)$$

It should be noted that, in determining Δ, the impedance Z_s must *not* be included in Z_{jj}, the impedance of the jth mesh.

5.5 OUTPUT IMPEDANCE

Consider a voltage source of emf V_s and internal impedance Z_s connected across the terminals AB of the jth mesh of a linear, n-mesh network, such that it supplies energy to a load of impedance Z_L connected across the terminals CD of the kth mesh, as shown in Fig. 5.5(a). The *output* impedance Z_k^{OUT} of this circuit is defined as the impedance across the terminals CD when Z_L is removed and all voltage sources in the n-mesh network have been replaced by their internal impedances. Thus, if a current I_k in the kth mesh is produced by a voltage V_k placed across the terminals CD, as shown in Fig. 5.5(b), then

$$Z_k^{OUT} = \frac{V_k}{I_k} \qquad (5.17)$$

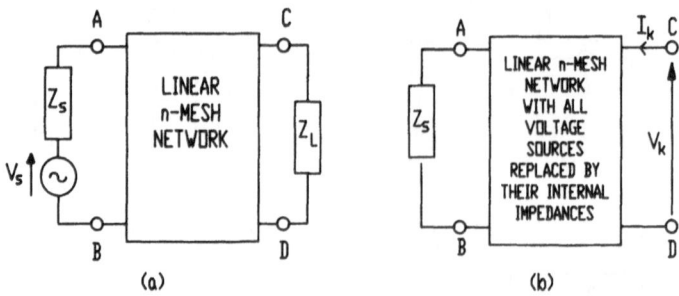

Fig. 5.5 (a) Linear n-mesh network with a voltage source connected in the jth mesh and a load connected in the kth mesh, and (b) circuit used to evaluate the output impedance, Z_k^{OUT}.

However, since all external voltage sources in the n-mesh network, except V_k, have been replaced by their internal impedances, (5.11) gives

$$I_k = \frac{\Delta_{kk}}{\Delta} V_k$$

which by comparison with (5.17) gives

$$Z_k^{OUT} = \frac{\Delta}{\Delta_{kk}} \tag{5.18}$$

It should be noted that, in determining Δ and Δ_{kk}, Z_s forms part of the jth mesh and must therefore be included in Z_{jj}, but Z_L must *not* be included in Z_{kk}.

5.6 TRANSFER IMPEDANCE

A linear, n-mesh network, as described in section 5.2, in which the only voltage source is V_k in the kth mesh will, from (5.11), have a current in the jth mesh given by

$$I_j = \frac{\Delta_{kj}}{\Delta} V_k$$

The *network transfer impedance*, Z_{kj}^{TRAN}, between kth and jth meshes is defined as the ratio of V_k to I_j; hence

$$Z_{kj}^{TRAN} = \frac{\Delta}{\Delta_{kj}} \tag{5.19}$$

It follows, therefore, if the emfs in the other meshes are not zero, the general expression for I_j will be

$$I_j = \sum_{k=1}^{n} V_k/Z_{kj}^{\text{TRAN}} \qquad (5.20)$$

However, it should be noted that in (5.20), when $k = j$

$$Z_{jj}^{\text{TRAN}} = Z_j^{\text{IN}}$$

In addition, the symmetry of the impedance matrix Δ given by (5.9) means that

$$Z_{kj}^{\text{TRAN}} = Z_{jk}^{\text{TRAN}}$$

The transfer impedance will increase, of course, as the separation of the two meshes, j and k, within the network increases.

5.7 NODAL ANALYSIS

Nodal analysis can be applied to both dc and ac circuits with the passive elements expressed as conductances or admittances and the power sources as current generators rather than voltage generators. As will be seen, compared with mesh analysis, this procedure will normally reduce the number of equations to be solved. The method is best introduced by an example such as the circuit shown in Fig. 5.6.

Voltages, which are to be determined, are assigned to each node as shown and one, in this case V_D, is chosen as reference and set equal to zero: this reduces the number of equations to be solved. Kirchhoff's current law applied to each node requires the algebraic

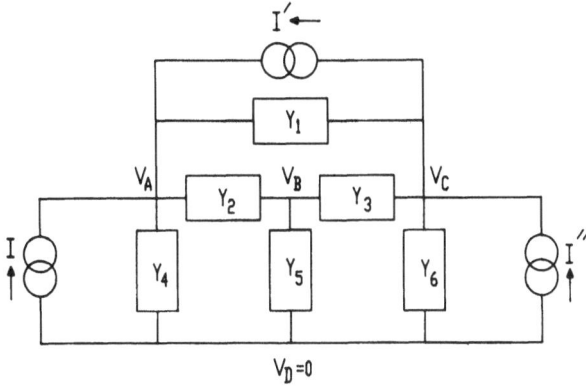

Fig. 5.6 The allocation of the node voltages to a circuit.

sum of the currents flowing into the node to be zero; thus a set of simultaneous equations for V_A, V_B and V_C can be derived as follows.

1. For node A

$$I + I' - (V_A - V_C)Y_1 - (V_A - V_B)Y_2 - V_A Y_4 = 0$$

where $(V_A - V_C)Y_1$ is the current flowing from node A to node C through Y_1, etc. Thus

$$(Y_1 + Y_2 + Y_4)V_A - Y_2 V_B - Y_1 V_C = I + I' \qquad (5.21(a))$$

2. For node B

$$- (V_B - V_A)Y_2 - (V_B - V_C)Y_3 - V_B Y_5 = 0$$

or

$$- Y_2 V_A + (Y_2 + Y_3 + Y_5)V_B - Y_3 V_C = 0 \qquad (5.21(b))$$

3. For node C

$$(I'' - I') - (V_C - V_A)Y_1 - (V_C - V_B)Y_3 - V_C Y_6 = 0$$

or

$$- Y_1 V_A - Y_3 V_B + (Y_1 + Y_3 + Y_6)V_C = I'' - I' \qquad (5.21(c))$$

Equations (5.21(a)), (5.21(b)) and (5.21(c)) show that the nodal equations can be obtained more readily by applying the following rule:

> for a particular node, the product of the node voltage and the sum of the admittances connected to it minus the sum of the products of the voltage of each adjacent node and the admittance connecting it to the particular node is equal to the sum of the currents flowing into the particular node from the external current generators.

The application of the above rule to the circuit shown in Fig. 5.6 enables the resulting nodal equations to be expressed directly in matrix form:

$$\begin{bmatrix} Y_1 + Y_2 + Y_4 & -Y_2 & -Y_1 \\ -Y_2 & Y_2 + Y_3 + Y_5 & -Y_3 \\ -Y_1 & -Y_3 & Y_1 + Y_3 + Y_6 \end{bmatrix} \begin{bmatrix} V_A \\ V_B \\ V_C \end{bmatrix} = \begin{bmatrix} I + I' \\ 0 \\ I'' - I' \end{bmatrix}$$

$$(5.22)$$

V_A, V_B and V_C can be determined by solving the matrix.

5.8 NODAL ANALYSIS APPLIED TO A GENERALIZED CIRCUIT

The generalized form of the matrix for a network having $n + 1$ nodes can be written as

$$\begin{bmatrix} Y_{11} & Y_{12} & \cdots & Y_{1k} & \cdots & Y_{1n} \\ Y_{21} & Y_{22} & \cdots & Y_{2k} & \cdots & Y_{2n} \\ \vdots & \vdots & & \vdots & & \vdots \\ Y_{j1} & Y_{j2} & \cdots & Y_{jk} & \cdots & Y_{jn} \\ \vdots & \vdots & & \vdots & & \vdots \\ Y_{n1} & Y_{n2} & \cdots & Y_{nk} & \cdots & Y_{nn} \end{bmatrix} \begin{bmatrix} V_1 \\ V_2 \\ \vdots \\ V_j \\ \vdots \\ V_n \end{bmatrix} = \begin{bmatrix} I_1 \\ I_2 \\ \vdots \\ I_j \\ \vdots \\ I_n \end{bmatrix} \quad (5.23(\text{a}))$$

or

$$[Y][V] = [I] \quad (5.23(\text{b}))$$

where V_j ($j = 1$ to n) is the potential of the jth node referred to the $(n + 1)$th node. I_j ($j = 1$ to n) is the algebraic sum of the currents flowing into the jth node from external current generators. Y_{jj} ($j = 1$ to n) is the sum of the admittances connected directly to the jth node. Y_{jk} for $j \neq k$ (j, $k = 1$ to n), is the *negative* sum of the admittances directly connecting the jth node to the kth node.

Let the determinant of the admittance matrix $[Y]$ be Δ'. Then, since the solution of (5.23(b)) is $[V] = [Y]^{-1}[I]$, it is possible to write

$$\begin{bmatrix} V_1 \\ V_2 \\ \vdots \\ V_j \\ \vdots \\ V_n \end{bmatrix} = \frac{1}{\Delta'} \begin{bmatrix} \Delta'_{11} & \Delta'_{21} & \cdots & \Delta'_{k1} & \cdots & \Delta'_{n1} \\ \Delta'_{12} & \Delta'_{22} & \cdots & \Delta'_{k2} & \cdots & \Delta'_{n2} \\ \vdots & \vdots & & \vdots & & \vdots \\ \Delta'_{1j} & \Delta'_{2j} & \cdots & \Delta'_{kj} & \cdots & \Delta'_{nj} \\ \vdots & \vdots & & \vdots & & \vdots \\ \Delta'_{1n} & \Delta'_{2n} & \cdots & \Delta'_{kn} & \cdots & \Delta'_{nn} \end{bmatrix} \begin{bmatrix} I_1 \\ I_2 \\ \vdots \\ I_j \\ \vdots \\ I_n \end{bmatrix} \quad (5.24)$$

where Δ'_{kj} represents the cofactor of Y_{kj} in the determinant Δ', as described earlier in section 5.2.

5.9 INPUT, OUTPUT AND TRANSFER ADMITTANCES

If the *input admittance* Y_j^{IN} at the jth node of a network of $n + 1$ nodes is defined as the ratio of the current I_j flowing into the network at the jth node from an external current source to the potential V_j acting between the jth node and the $(n + 1)$th reference node when all other current sources are zero, then from (5.24)

$$V_j = \frac{\Delta'_{jj}}{\Delta'} I_j \quad \text{and} \quad Y_j^{IN} = \frac{\Delta'}{\Delta'_{jj}} \tag{5.25}$$

The situation is illustrated in Fig. 5.7 and it should be noted that, when determining Δ', the internal admittance Y_s of the current source I_s must *not* be included in the sum of admittances connected to the jth node.

Consider now the circuit shown in Fig. 5.8(a) in which an external current source I_s, of internal admittance Y_s, is connected between the eth and fth nodes of a network containing $n+1$ nodes, in order to supply energy to a load of admittance Y_L connected between the kth node and the $(n+1)$th reference node. The *output admittance*, Y_k^{OUT}, of the network at the kth node is defined as the admittance between the kth node and the $(n+1)$th reference node when Y_L has been removed and all current sources, such as I_s, have been reduced to zero. Thus, if a current I_k flows into the kth node when a voltage V_k is placed across the kth and $(n+1)$th nodes, as shown in Fig. 5.8(b), then, from (5.24),

$$V_k = \frac{\Delta'_{kk}}{\Delta'} I_k \quad \text{and} \quad Y_k^{OUT} = \frac{\Delta'}{\Delta'_{kk}} \tag{5.26}$$

It should be noted that, when evaluating Δ' and Δ'_{kk}, the admittance, Y_s, of the current source, I_s, must be included in the terms Y_{ee}, Y_{ef}, Y_{fe} and Y_{ff}, but the load admittance, Y_L, must not be included in Y_{kk}.

If, as shown in Fig. 5.9. a pd V_j is produced between the jth node and the $(n+1)$th reference node of a linear network containing $n+1$ nodes, when a current I_k flows into the kth node from a current

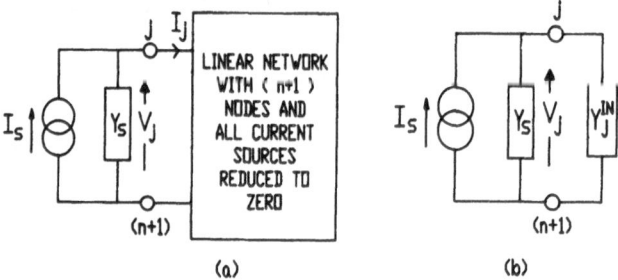

Fig. 5.7 (a) Current source applied between node j and the reference node $n+1$ of a linear network with $n+1$ nodes, and (b) the equivalent circuit.

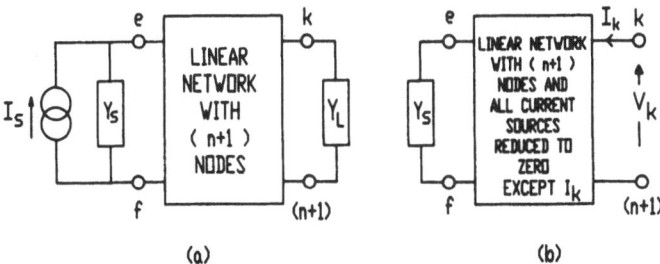

Fig. 5.8 (a) Linear network with $n + 1$ nodes and a current source applied between the eth and fth nodes: a load admittance, Y_L, is connected between the kth and $(n + 1)$th nodes, and (b) circuit used to evaluate the output admittance, Y_k^{OUT}.

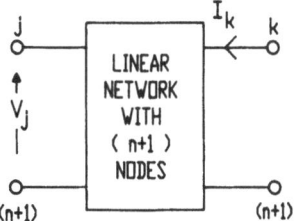

Fig. 5.9 Current I_k flowing into the kth node of a linear network with $n + 1$ nodes, thus causing a pd V_j to appear between the jth node and the reference node $n + 1$.

source, then from (5.24)

$$V_j = \frac{\Delta'_{kj}}{\Delta'} I_k$$

The *network transfer admittance*, Y_{kj}^{TRAN}, between the kth and jth nodes is defined as the ratio of the current I_k to the pd V_j and hence

$$Y_{kj}^{TRAN} = \frac{\Delta'}{\Delta'_{kj}} \tag{5.27}$$

When more than one current source is connected to the network then, from (5.24), the general expression for V_j becomes

$$V_j = \sum_{k=1}^{n} (I_k / Y_{kj}^{TRAN}) \tag{5.28}$$

However, in (5.28), it should be noted that, when $k = j$,

$$Y_{jj}^{TRAN} = Y_j^{IN}$$

In addition, the symmetry of the admittance matrix Δ' given by (5.23) means that

$$Y_{kj}^{\text{TRAN}} = Y_{jk}^{\text{TRAN}}$$

The transfer admittance will increase, of course, as the separation of the nodes, j and k, within the network increases.

Since it is often impossible to devise an appropriate, simple method for designating the nodes, the admittances defined as Y_j^{IN}, Y_k^{OUT} and Y_{kj}^{TRAN} have somewhat more restricted applications than the corresponding impedances. Nevertheless, the method of nodal analysis can be used as a powerful tool in the analysis of electrical networks.

6

Network theorems and transformations

By applying the techniques of either mesh or nodal analysis it is possible to solve any problem involving the relationships between the currents and voltages in a network. However, in many cases, the solution of such problems can be simplified by the use of one or more of the theorems discussed in this chapter and which are applicable to both ac and dc linear networks.

6.1 THÉVENIN'S THEOREM

Thévenin's theorem states that

> any two-terminal, linear network of voltage generators and impedances can be replaced by a single voltage generator in series with a single impedance.

To prove this theorem and to identify the single voltage generator and single impedance, consider the generalized impedance network

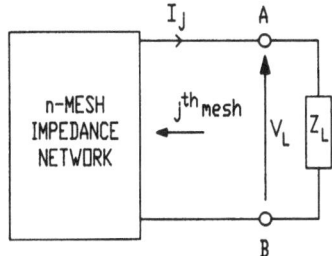

Fig. 6.1 An n-mesh impedance network with a load Z_L connected into the jth mesh.

discussed in section 5.2 which consisted of n meshes of impedances and voltage generators. Suppose that the jth mesh of this network is broken so that the network now has two accessible output terminals A and B as shown in Fig. 6.1 and let a load of impedance Z_L be connected across these two terminals. The current I_j flowing in the jth mesh is given by (5.11), whence it will be seen that

$$I_j = \frac{\Delta_j}{\Delta'} = \frac{\Delta_{1j}}{\Delta'} V_1 + \frac{\Delta_{2j}}{\Delta'} V_2 + \cdots + \frac{\Delta_{nj}}{\Delta'} V_n \qquad (6.1)$$

where Δ, which includes the impedance Z_{jj}, has been replaced by Δ' in which Z_{jj} is replaced by $Z_{jj} + Z_L$. Hence

$$\Delta' = \begin{bmatrix} Z_{11} & Z_{12} & \cdots & Z_{1j} & \cdots & Z_{1n} \\ Z_{21} & Z_{22} & \cdots & Z_{2j} & \cdots & Z_{2n} \\ \vdots & \vdots & & \vdots & & \vdots \\ Z_{j1} & Z_{j2} & \cdots & Z_{jj} + Z_L & \cdots & Z_{jn} \\ \vdots & \vdots & & \vdots & & \vdots \\ Z_{n1} & Z_{n2} & \cdots & Z_{nj} & \cdots & Z_{nn} \end{bmatrix}$$

$$= \Delta + Z_L \Delta_{jj}$$

where Δ_{jj} is the cofactor of Z_{jj}.

Substituting for Δ' in (6.1) gives

$$I_j = \frac{1}{\Delta + Z_L \Delta_{jj}} (\Delta_{1j} V_1 + \Delta_{2j} V_2 + \cdots + \Delta_{nj} V_n) \qquad (6.2)$$

The voltage V_L, across the load, is given by

$$V_L = I_j Z_L = \frac{Z_L}{\Delta + Z_L \Delta_{jj}} (\Delta_{1j} V_1 + \Delta_{2j} V_2 + \cdots + \Delta_{nj} V_n)$$

$$= \frac{1/\Delta_{jj}}{1 + \Delta/Z_L \Delta_{jj}} \sum_{k=1}^{n} \Delta_{kj} V_k$$

In the open-circuit case, $Z_L = \infty$ and $V_L = V_{oc}$, the open-circuit voltage, and so

$$V_{oc} = \frac{1}{\Delta_{jj}} \sum_{k=1}^{n} \Delta_{kj} V_k$$

Substituting in (6.2) and rearranging gives

$$I_j = \frac{V_{oc}}{Z_L + \Delta/\Delta_{jj}} \qquad (6.3)$$

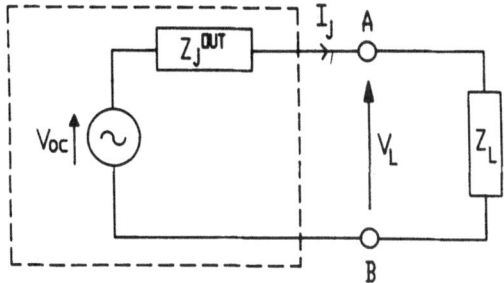

Fig. 6.2 Thévenin's equivalent voltage generator.

But, from (5.18), Δ/Δ_{jj} is the output impedance, Z_j^{OUT}, of the network, hence

$$I_j = \frac{V_{oc}}{Z_j^{OUT} + Z_L} \tag{6.4}$$

Equation (6.4) represents a mathematical statement of Thévenin's theorem and Thévenin's equivalent voltage generator can be represented as in Fig. 6.2.

When deriving Thévenin's equivalent voltage generator it must be remembered that (a) V_{oc} is the voltage measured between the terminals A and B of the network when no load or external circuit is connected across them, and (b) Z_j^{OUT} is the impedance measured between the terminals when all sources of emf have been replaced by a short circuit and all current sources have been removed.

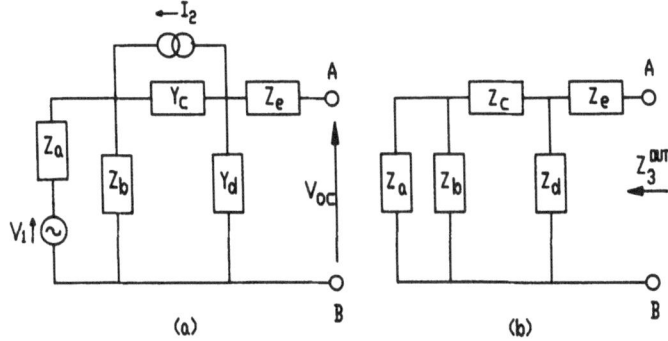

Fig. 6.3 (a) Circuit consisting of three meshes and having both voltage and current sources. (b) Equivalent circuit used to determine Z_3^{OUT} in which $Z_c = Y_c^{-1}$ and $Z_d = Y_d^{-1}$.

The process whereby the output impedance, Z_j^{OUT}, can be determined for a network consisting of three meshes is shown in Figs 6.3(a) and 6.3(b). By reference to (5.7), it will be seen that the impedance matrix for the circuit shown in Fig. 6.3(b) gives

$$\Delta = \begin{bmatrix} Z_{11} & Z_{12} & Z_{13} \\ Z_{21} & Z_{22} & Z_{23} \\ Z_{31} & Z_{32} & Z_{33} \end{bmatrix} = \begin{bmatrix} Z_a + Z_b & -Z_b & 0 \\ -Z_b & Z_b + Z_c + Z_d & -Z_d \\ 0 & -Z_d & Z_d + Z_e \end{bmatrix}$$

and $Z_3^{OUT} = \Delta/\Delta_{33}$ where Δ_{33} is the cofactor of Z_{33}. It is obvious that, for simple networks, it is easier to calculate Z_3^{OUT} by directly considering the series and parallel arrangements of impedances which form the network.

The reduction of a two-terminal network of generators and impedances to a single equivalent two-terminal voltage generator is particularly useful when it is required to investigate the effect of varying the load connected across the two terminals of the network.

6.2 NORTON'S THEOREM

Norton's theorem states that

> any two-terminal, linear network of current generators and admittances can be replaced by a single current generator in parallel with a single admittance.

The proof of this theorem and the identification of the single current generator and single admittance can be achieved by considering the generalized admittance network discussed in section 5.8. This network has $n + 1$ nodes and consists of current generators and admittances. Let the network also have two accessible output terminals connected to node j and node $n + 1$ and let a load of

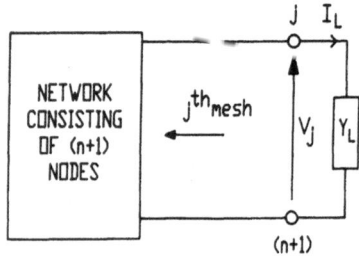

.**Fig. 6.4** Network consisting of $n + 1$ nodes with a load of admittance Y_L connected between nodes j and $n + 1$.

admittance Y_L, be connected across the output terminals as shown in Fig. 6.4.

The potential V_j of node j relative to the node $n + 1$, which is assumed to be at zero potential, is given by (5.24), whence

$$V_j = \frac{\Delta'_{1j}}{\Delta''} I_1 + \frac{\Delta'_{2j}}{\Delta''} I_2 + \cdots + \frac{\Delta'_{nj}}{\Delta''} I_n \qquad (6.5)$$

where Δ', which includes the admittance Y_{jj}, has been replaced by Δ'' in which Y_{jj} is replaced by $Y_{jj} + Y_L$, so that

$$\Delta'' = \begin{bmatrix} Y_{11} & Y_{12} & \cdots & Y_{1j} & \cdots & Y_{1n} \\ Y_{21} & Y_{22} & \cdots & Y_{2j} & \cdots & Y_{2n} \\ \vdots & \vdots & & \vdots & & \vdots \\ Y_{j1} & Y_{j2} & \cdots & Y_{jj} + Y_L & \cdots & Y_{jn} \\ \vdots & \vdots & & \vdots & & \vdots \\ Y_{n1} & Y_{n2} & \cdots & Y_{nj} & \cdots & Y_{nn} \end{bmatrix}$$

where Δ'_{jj} is the cofactor of Y_{jj}. Substituting for Δ'' in (6.5) gives

$$V_j = \frac{1}{\Delta' + Y_L \Delta'_{jj}} (\Delta'_{1j} I_1 + \Delta'_{2j} I_2 + \cdots + \Delta'_{nj} I_n)$$

The currentt I_L flowing through the load Y_L is given by

$$I_L = Y_L V_j = \frac{1}{\Delta'_{jj} + \Delta' + Y_L \Delta'_{jj}/Y_L} \sum_{k=1}^{n} \Delta'_{kj} I_k \qquad (6.6)$$

$$\equiv \Delta' + Y_L \Delta'_{jj}$$

If the output terminals of the network are short circuited, then $Y_L \to \infty$ and $I_L \to I_{sc}$, the short-circuited current, and so

$$I_{SC} = \frac{1}{\Delta'_{jj}} \sum_{k=1}^{n} \Delta'_{kj} I_k$$

Substituting in (6.6) and rearranging gives

$$I_L = \frac{Y_L \Delta'_{jj}}{\Delta' + Y_L \Delta'_{jj}} I_{SC} = \frac{Y_L}{Y_L + \Delta'/\Delta'_{jj}} I_{sc} \qquad (6.7)$$

But, from (5.26), Δ'/Δ'_{jj} corresponds to the output admittance at the jth node, Y_j^{OUT}; hence

$$I_L = \frac{Y_L I_{sc}}{Y_j^{OUT} + Y_L} \qquad (6.8)$$

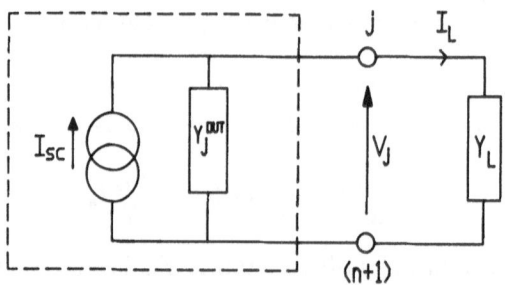

Fig. 6.5 Norton's equivalent current generator with load Y_L.

Equation (6.8) represents a mathematical statement of Norton's theorem and Norton's equivalent current generator is shown in Fig. 6.5, inside the broken rectangle.

As in the case of Thévenin's theorem, Norton's theorem is useful for investigating the effect of varying the load connected across the two output terminals of the network.

It should be noted that Thévenin's voltage generator can be converted directly into the equivalent Norton's current generator, and vice versa, since

$$Z_j^{OUT} = \frac{\Delta}{\Delta_{jj}} = \frac{1}{Y_j^{OUT}} = \frac{\Delta'_{jj}}{\Delta'}$$

and

$$V_{oc} = I_{sc} Z_j^{OUT} \quad \text{or} \quad I_{sc} = V_{oc} Y_j^{OUT} \tag{6.9}$$

When deriving Norton's equivalent current generator it must be remembered that (a) I_{SC} is the current that would flow through a short circuit placed between the output terminals of the network, and (b) V_j^{OUT} is the admittance measured between the output terminals when all sources of emf have been replaced by a short circuit and all current sources have been removed. The process whereby the output admittance Y_j^{OUT} can be determined for a network having four nodes is shown in Figs 6.6(a) and 6.6(b). By reference to (5.23(a)), it will be seen that the admittance matrix for the circuit shown in Fig. 6.6(b) gives

$$\Delta' = \begin{bmatrix} Y_{11} & Y_{12} & Y_{13} \\ Y_{21} & Y_{22} & Y_{23} \\ Y_{31} & Y_{32} & Y_{33} \end{bmatrix} = \begin{bmatrix} Y_a + Y_b + Y_c & -Y_c & 0 \\ -Y_c & Y_c + Y_d + Y_e & -Y_e \\ 0 & -Y_e & Y_e \end{bmatrix}$$

and $Y_3^{OUT} = \Delta'/\Delta'_{33}$ where Δ'_{33} is the cofactor of Y_{33}. However, it is obvious that, for simple circuits, it will usually be easier to calculate

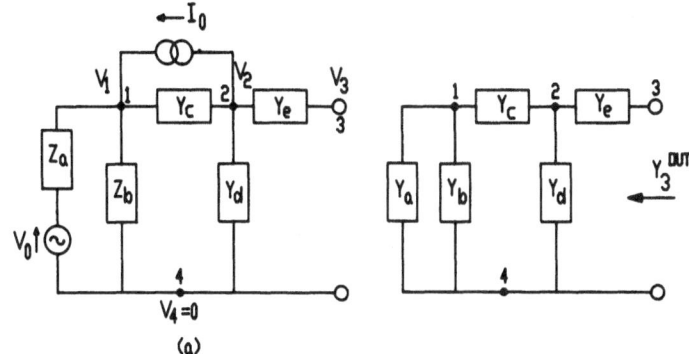

Fig. 6.6 (a) Circuit with four nodes and having both voltage and current sources. (b) Equivalent circuit used to determine Y_3^{OUT} in which $Y_a = Z_a^{-1}$ and $Y_b = Z_b^{-1}$.

Y_3^{OUT} by directly considering the series and parallel arrangement of admittances which form the network.

6.3 MILLMAN'S THEOREM

Millman's theorem states that

> a network which consists of either a number of voltage generators connected in parallel or a number of current generators connected in series can be reduced to either a single current generator or a single voltage generator.

To prove this theorem, first consider a circuit of n parallel branches; and let each branch be connected between nodes 1 and 2 and consist of a voltage source of constant emf in series with an impedance as shown in Fig. 6.7(a). Each branch can be converted into an equivalent current source as shown in Fig. 6.7(b) such that $I_k = V_k/Z_k$ and $Y_k = Z_k^{-1}$ (see, e.g. section 2.2.3); these current sources can then be reduced to the equivalent Norton's current generator (I_{SC}, Y_1^{OUT}), and hence to the corresponding Thévenin's voltage generator (V_{oc}, Z_1^{OUT}), as shown in Fig. 6.7(c), where

$$I_{sc} = \sum_{k=1}^{n} I_k = \sum_{k=1}^{n} V_k(Z_k)^{-1} \tag{6.10}$$

$$Y_1^{OUT} = \sum_{k=1}^{n} Y_k = \sum_{k=1}^{n} (Z_k)^{-1}$$

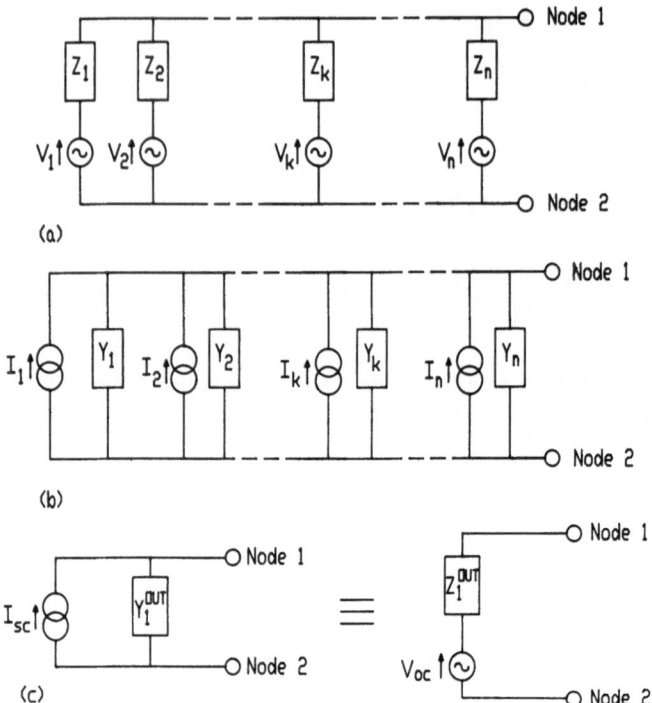

Fig. 6.7 (a) *n* sets of parallel branches each consisting of a voltage source in series with an impedance. (b) Corresponding set of *n* parallel current sources. (c) Equivalent Norton's current generator and Thévenin's voltage generator.

and

$$V_{oc} = I_{sc}(Y_1^{OUT})^{-1} = \left[\sum_{k=1}^{n} V_k(Z_k)^{-1} \right] \left[\sum_{k=1}^{n} (Z_k)^{-1} \right]^{-1}$$

$$Z_1^{OUT} = (Y_1^{OUT})^{-1} = \left[\sum_{k=1}^{n} (Z_k)^{-1} \right]^{-1}$$

(6.11)

In the case of the circuit consisting of *n* current generators connected in series, where each current generator consists of a source of constant current in parallel with an admittance as in Fig. 6.8(a), each current generator can be converted into an equivalent voltage generator as shown in Fig. 6.8(b), so that $V_k = I_k(Y_k)^{-1}$ and

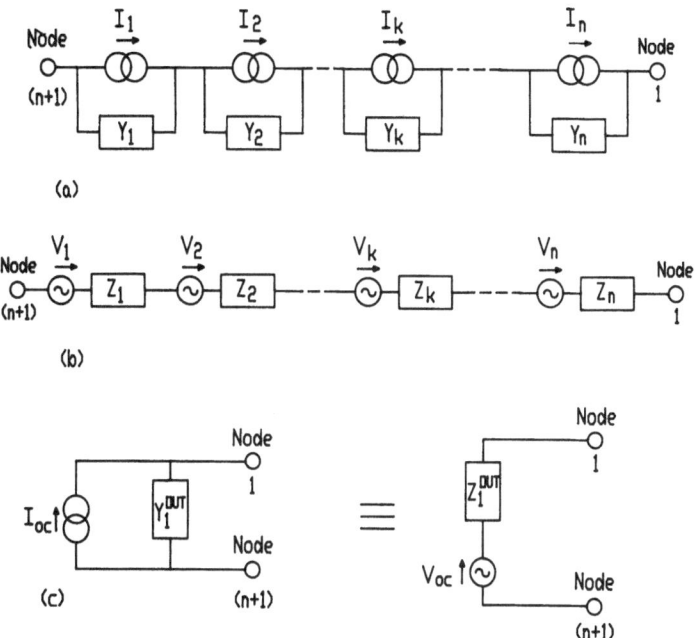

(a)

(b)

(c)

Fig. 6.8 (a) n current generators connected in series. (b) n corresponding voltage generators connected in series. (c) Equivalent Thévenin's voltage generator and the corresponding Norton's current generator.

$Z_k = (Y_k)^{-1}$. In turn, this new circuit reduces to the equivalent Thévenin's voltage generator (V_{oc}, Z_1^{OUT}), and hence to the corresponding Norton's current generator (I_{sc}, Y_1^{OUT}), as shown in Fig. 6.8(c), where

$$V_{oc} = \sum_{k=1}^{n} V_k = \sum_{k=1}^{n} I_k(Y_k)^{-1}$$

$$Z_1^{OUT} = \sum_{k=1}^{n} Z_k = \sum_{k=1}^{n} (Y_k)^{-1}$$

(6.12)

and

$$I_{SC} = V_{oc}/Z_1^{OUT} = \left[\sum_{k=1}^{n} I_k(Y_k)^{-1} \right]\left[\sum_{k=1}^{n} (Y_k)^{-1} \right]^{-1}$$

$$Y_1^{OUT} = (Z_1^{OUT})^{-1} = \left[\sum_{k=1}^{n} (Y_k)^{-1} \right]^{-1}$$

(6.13)

The successive application of Millman's theorem to a network which consists of a number of branches containing voltage and/or current generators, impedances and/or admittances enables that network to be reduced to an equivalent single generator. Hence, Millman's theorem is extremely useful when determining either the Thévenin's voltage generator or the Norton's current generator which is equivalent to a given multi-source network.

6.4 THE RECIPROCITY THEOREM

The reciprocity theorem can be stated in two forms as follows: either

1. in a linear network, if a voltage source placed in the kth mesh produces a certain current in the jth mesh, then the same voltage source placed in the jth mesh will produce the same current in the kth mesh, or
2. in a linear network, if a current flowing from an external source into the kth node produces a certain voltage at the jth node, then the same current flowing into the jth node will produce the same voltage at the kth node.

In order to verify the first statement of the reciprocity theorem, consider a linear, n-mesh network, as described in section 5.2, in which the only voltage source is V_k in the kth mesh. The current I_j in the jth mesh is given by (5.20), whence

$$I_j = V_k / Z_{kj}^{\text{TRAN}}$$

where Z_{kj}^{TRAN} is the network transfer impedance between the kth and jth meshes as defined in section 5.6.

Similarly, if the only source is V_j in the jth mesh, the current in the kth mesh is

$$I_k = V_j / Z_{jk}^{\text{TRAN}}$$

But from the symmetry of the impedance matrix, $Z_{kj}^{\text{TRAN}} = Z_{jk}^{\text{TRAN}}$; hence, if $V_k = V_j$, then $I_j = I_k$.

The second statement of the reciprocity theorem can be verified by considering a linear network which contains $n + 1$ nodes, as described in section 5.8, in which the only current source is the current I_k flowing into the kth node. This current will produce a voltage V_j at the jth node given by (5.28), i.e.

$$V_j = (Y_{kj}^{\text{TRAN}})^{-1} I_k$$

where Y_{kj}^{TRAN} is the network transfer impedance between the kth and jth nodes as defined in section 5.9.

Similarly, if the only current source is the current I_j flowing into the jth node, then the voltage V_k produced at the kth node will be

$$V_k = (Y_{jk}^{\text{TRAN}})^{-1} I_j$$

From the symmetry of the admittance matrix, $Y_{jk}^{\text{TRAN}} = Y_{kj}^{\text{TRAN}}$; and, if $I_k = I_j$, then $V_j = V_k$.

6.5 THE SUPERPOSITION THEOREM

The superposition theorem can be stated in two forms as follows: either

1. in any linear network of impedances and voltage sources, the current flowing in a given mesh is equal to the sum of the currents flowing in that mesh due to each voltage source taken separately, when all other voltage sources have been replaced by a short circuit, or
2. in any linear network of admittances and current sources, the potential at a given node is equal to the sum of the potentials at that node due to each current source taken separately, when all other current sources have been removed.

For a linear, n-mesh network of voltage generators and impedances, as described in section 5.2, the current in the jth mesh is given by (5.20), i.e.

$$I_j = \sum_{k=1}^{n} \frac{V_k}{Z_{kj}^{\text{TRAN}}}$$

This equation is simply a mathematical statement of the first form of the superposition theorem as stated above.

Similarly, for a linear network containing $n + 1$ nodes and consisting of current sources and admittances, as described in section 5.8, the voltage of the jth node is given by (5.28), i.e.

$$V_j = \sum_{k=1}^{n} I_k (Y_{kj}^{\text{TRAN}})^{-1}$$

and this equation represents a mathematical statement of the second form of the superposition theorem stated above.

The superposition theorem is useful for determining the effect in a particular part of a circuit, whether mesh or node, when the voltage or current source in another part of the circuit is altered. It should be noted, however, that superposition should not be applied in the case of power which is not linear in voltage or current but is proportional to the squares of these quantities.

6.6 THE SUBSTITUTION THEOREM

The substitution theorem really represents a statement of the obvious and can be quoted as follows:

if any part of a circuit is replaced by another combination of components such that the conditions at the output terminals are unchanged then the currents and voltages within the remainder of the circuit will be unaltered.

6.7 THE COMPENSATION THEOREM

The compensation theorem results from a combination of the substitution theorem and the superposition theorem and can be stated as follows:

if an impedance Z in a branch of a linear circuit carries a current I and the impedance is changed by an amount ΔZ, then the current flowing in another branch of the circuit will change by an amount equal to that produced in that branch when Z is replaced by a voltage generator of internal impedance $Z + \Delta Z$ and emf $V = I \Delta Z$, in opposition to I, and all other voltage sources in the circuit have been replaced by short circuits and all current sources have been removed.

The compensation theorem is often used in bridge or potentiometer circuits in order to determine the change from the null condition caused by a slight variation in the value of one of the impedances in the circuit. It is also used for estimating the errors caused by the introduction of, say, a meter into a circuit for the purpose of measuring a potential difference or a current.

6.8 THE STAR–DELTA (λ– Δ) TRANSFORMATION

A three-terminal network in the form of a delta, Δ, such as that shown in Fig. 6.9(a), can be changed into an equivalent three-

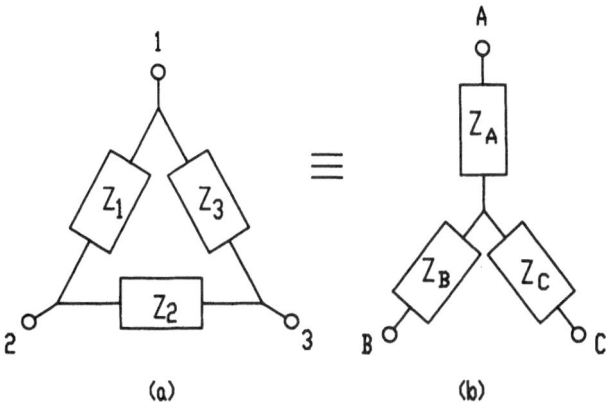

Fig. 6.9 (a) Three-terminal network in the form of a delta. (b) Equivalent three-terminal network in the form of a star.

terminal network in the form of a star,⅄, as shown in Fig. 6.9(b), and vice versa.

In order to transform the delta network of Fig. 6.9(a) into the equivalent star network of Fig. 6.9(b), it is necessary first to derive the three equations for the impedance between each pair of terminals when the third terminal is open circuited; then

$$Z_{AB} = Z_A + Z_B = \frac{Z_1(Z_2 + Z_3)}{Z_1 + Z_2 + Z_3}$$

$$Z_{BC} = Z_B + Z_C = \frac{Z_2(Z_3 + Z_1)}{Z_1 + Z_2 + Z_3} \quad (6.14)$$

$$Z_{CA} = Z_C + Z_A = \frac{Z_3(Z_1 + Z_2)}{Z_1 + Z_2 + Z_3}$$

The solution of these three simultaneous equations is

$$Z_A = \frac{Z_1 Z_3}{Z_1 + Z_2 + Z_3}$$

$$Z_B = \frac{Z_2 Z_1}{Z_1 + Z_2 + Z_3} \quad (6.15)$$

$$Z_C = \frac{Z_3 Z_2}{Z_1 + Z_2 + Z_3}$$

or, in terms of the branch admittances,

$$Y_A = \frac{Y_1 Y_2 + Y_2 Y_3 + Y_3 Y_1}{Y_2}$$

$$Y_B = \frac{Y_1 Y_2 + Y_2 Y_3 + Y_3 Y_1}{Y_3} \qquad (6.16)$$

$$Y_C = \frac{Y_1 Y_2 + Y_2 Y_3 + Y_3 Y_1}{Y_1}$$

Obviously, it is possible to transpose these two sets of equations so that a conversion can be made in the reverse direction, i.e. star to delta. However, it is much simpler to equate the admittances between two terminals when one of the other pairs of terminals is short circuited; thus

AB short circuited: $Y_{BC} = Y_2 + Y_3 = \dfrac{Y_C(Y_A + Y_B)}{Y_A + Y_B + Y_C}$

BC short circuited: $Y_{CA} = Y_3 + Y_1 = \dfrac{Y_A(Y_B + Y_C)}{Y_A + Y_B + Y_C}$ (6.17)

CA short circuited: $Y_{AB} = Y_1 + Y_2 = \dfrac{Y_B(Y_C + Y_A)}{Y_A + Y_B + Y_C}$

These three equations have the same form as (6.14); hence

$$Y_1 = \frac{Y_A Y_B}{Y_A + Y_B + Y_C}$$

$$Y_2 = \frac{Y_B Y_C}{Y_A + Y_B + Y_C} \qquad (6.18)$$

$$Y_3 = \frac{Y_C Y_A}{Y_A + Y_B + Y_C}$$

which, in terms of branch impedances, becomes

$$Z_1 = \frac{Z_A Z_B + Z_B Z_C + Z_C Z_A}{Z_C}$$

$$Z_2 = \frac{Z_A Z_B + Z_B Z_C + Z_C Z_A}{Z_A} \qquad (6.19)$$

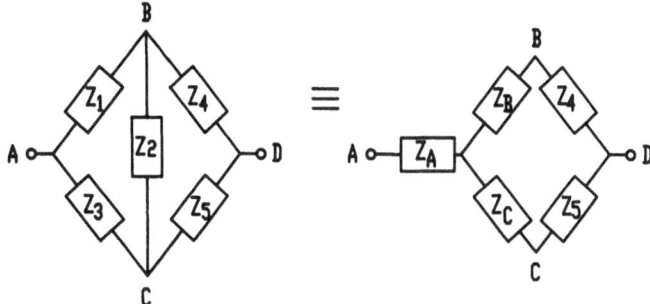

Fig. 6.10 Transformation of the delta ABC into the corresponding star ABC, thus enabling the impedance between the terminals A and D to be calculated.

$$Z_3 = \frac{Z_A Z_B + Z_B Z_C + Z_C Z_A}{Z_B}$$

When determining the effective impedance of a circuit, it is sometimes necessary to use the star–delta transformation in order to change the configuration of the circuit elements, so that the rules for impedances in series and parallel can then be applied. A simple example is illustrated in Fig. 6.10. Z_A, Z_B and Z_C can be obtained in terms of Z_1, Z_2 and Z_3 by using (6.15), and

$$Z_{AD} = Z_A + \frac{(Z_B + Z_4)(Z_C + Z_5)}{Z_B + Z_C + Z_4 + Z_5}$$

6.9 TRANSFORMATION INVOLVING A MUTUAL INDUCTANCE

The analysis of circuits coupled by a mutual inductance can often be simplified by transforming the mutual inductance into three self-inductances, arranged in the form of a T, with no mutual inductance existing between them. The conditions relating to such a transformation are derived below.

Figure 6.11(a) shows two circuits coupled by a mutual inductance M: the corresponding mesh equations are

$$V_1 = (Z_1 + j\omega L_1)I_1 \pm j\omega M I_2$$

and   (6.20)

$$0 = (Z_2 + j\omega L_2)I_2 \pm j\omega M I_1$$

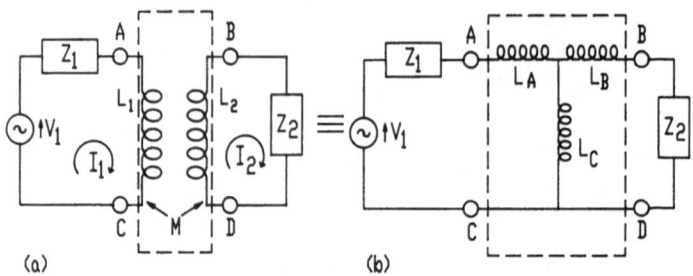

Fig. 6.11 (a) Two circuits coupled by a mutual inductance. (b) The equivalent circuit.

where the $\pm$ sign corresponds to the relative direction of the windings of the two coils in accordance with the dot notation which was introduced in section 5.3.

Similarly, the mesh equations corresponding to the equivalent circuit shown in Fig. 6.11(b), where the mutual inductance has been replaced by the three self-inductances L_A, L_B and L_C, are

$$V_1 = (Z_1 + j\omega L_A + j\omega L_C)I_1 - j\omega L_C I_2$$

and   (6.21)

$$0 = (Z_2 + j\omega L_B + j\omega L_C)I_2 - j\omega L_C I_1$$

The circuits shown in Figs 6.11(a) and 6.11(b) are equivalent only if (6.20) and (6.21) are equivalent, i.e.

$$L_1 = L_A + L_B, \pm M = -L_C, L_2 = L_B + L_C \qquad (6.22)$$

or

$$L_A = L_1 \pm M, L_B = L_2 + M, L_C = \pm M \qquad (6.23)$$

The derivation of the conditions represented in (6.23) assumes that the terminals C and D of the circuit shown in Fig. 6.11(a) may be regarded as being at the same potential.

The above transformation is particularly useful when determining the balance conditions for ac bridges which have been designed to measure mutual inductance, e.g. Maxwell's self- and mutual inductance bridge.

7

Electrical resonance

Most readers will be familiar with the phenomenon of resonance in acoustic or mechanical systems which occurs when the natural frequency of oscillation of the system coincides with that of the driving signal—a musical note or a vibration from which the system can absorb energy. Such a system will also absorb energy from the driving signal when the frequency of oscillation of the signal is very different from the natural frequency, but the amount of energy absorbed will be relatively small. As the driver frequency approaches the natural frequency, the amount of energy absorbed will increase, becoming a maximum when, if there is no damping in the system, the two frequencies are equal. The frequency at which maximum energy is absorbed is known as the *resonant* frequency and, when it is driven at this frequency, the system is said to *resonate* with the applied signal.

The phenomena of resonance can also occur in those electrical circuits in which both electrical and magnetic energy can be stored, i.e. circuits which contain both capacitive and inductive elements. The ac response of some simple circuits is considered in this chapter together with their more important general characteristics.

7.1 THE SERIES L-C-R CIRCUIT

If the circuit shown in Fig. 7.1 is driven at an angular frequency ω, its impedance will be given by

$$Z = R + j\omega L + \frac{1}{j\omega C} = R + j\left(\omega L - \frac{1}{\omega C}\right) \qquad (7.1(a))$$

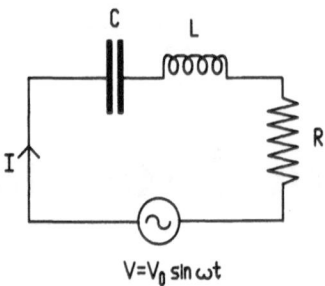

Fig. 7.1 Series L–C–R circuit.

and

$$|Z| = \left[R^2 + \left(\omega L - \frac{1}{\omega c} \right)^2 \right]^{1/2}$$
(7.1(b))

If the angular frequency is adjusted to ω_0, such that $\omega_0 L = 1/\omega_0 c$, then $|Z|$ is a minimum, and Z is real and equal to R. This condition corresponds to *resonance* in the circuit when

$$\omega_0 = (LC)^{-1/2} \quad \text{or} \quad f_0 = \frac{1}{2\pi\sqrt{LC}}$$
(7.2)

where f_0 is the *resonant frequency*. It should be noted that the resonant frequency is not the same as the *natural* frequency of the circuit, given by (3.26) as $f = (1/2\pi)(1/LC - R^2/4L^2)^{1/2}$, unless $R = 0$.

At resonance, the current is in phase with the applied voltage and has a maximum value I_m given by

$$I_m = V/R$$
(7.3)

The potential across the resistance R is equal to the applied voltage whilst the potentials across the inductor and capacitor are equal, are $180°$ out of phase with each other, and cancel.

Combining (7.1(a)) and (7.2) gives

$$Z = R + j\left(\omega L - \frac{\omega_0^2 L}{\omega} \right) = R\left[1 + \frac{j\omega_0 L}{R}\left(\frac{\omega}{\omega_0} - \frac{\omega_0}{\omega} \right) \right]$$

or

$$Z = R(1 + jQx)$$
(7.4)

where

$$Q = \frac{\omega_0 L}{R} = \frac{1}{\omega_0 CR} = \frac{1}{R}\sqrt{\frac{L}{C}}$$

is known as the *quality factor*, or more simply the *Q*, of the circuit and

$$x = \frac{\omega}{\omega_0} - \frac{\omega_0}{\omega} = \frac{f}{f_0} - \frac{f_0}{f} \tag{7.5}$$

From (7.4)

$$|Z| = R(1 + Q^2 x^2)^{1/2} \tag{7.6}$$

If the voltage amplitude is kept constant while the frequency is varied, then the ratio of the current in the circuit to the resonant current is

$$\frac{|I|}{I_m} = \frac{V}{|Z|} \bigg/ \frac{V}{R} = \frac{R}{|Z|} = \left[1 + Q^2 \left(\frac{f}{f_0} - \frac{f_0}{f} \right)^2 \right]^{-1/2} \tag{7.7}$$

The form of (7.7) is universal, depending on parameters Q and f_0 as shown in Fig. 7.2. Clearly the circuit is selective in its response to signals of different frequency, the selectivity increasing with Q.

When $|I|/I_m = 1/\sqrt{2}$, the power dissipated in the circuit is reduced by a factor 2 and the corresponding frequencies are known as the

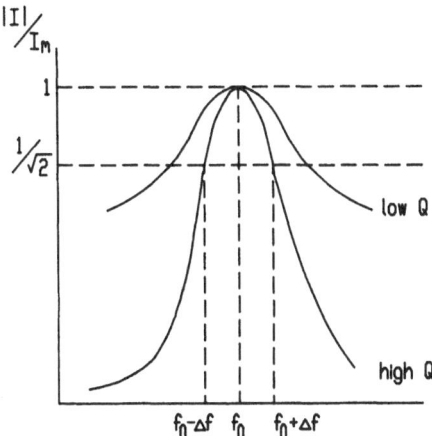

Fig. 7.2 The normalized response curves of a series $L–C–R$ circuit.

half-power points. The half-power points are obtained from (7.7) and correspond to the condition that

$$Q^2\left(\frac{f}{f_0}-\frac{f_0}{f}\right)^2 = 1 \quad \text{or} \quad \frac{f}{f_0}-\frac{f_0}{f} = \pm\frac{1}{Q}$$

If f is written as $f_0 \pm \Delta f$, where $\Delta f \ll f_0$,

$$\frac{f}{f_0}-\frac{f_0}{f}=\frac{f^2-f_0^2}{ff_0}=\frac{(f+f_0)(f-f_0)}{ff_0}\simeq\frac{2f_0(f-f_0)}{ff_0}=\pm\frac{\Delta f}{f_0}$$

Hence,

$$\Delta f=\frac{f_0}{2Q} \quad \text{or} \quad Q=\frac{f_0}{2\,\Delta f} \qquad (7.8)$$

The separation of the half-power points, $2\Delta f$, is known as the *linewidth.*

It should be noted that the resonance curve can only be treated as symmetrical for high values of Q, small values of $\Delta f/f_0$, yet the linewidth is still given by (7.8) for any value of Q as will now be shown.

The condition for $|I|/I_m = 1/\sqrt{2}$ is that

$$|Z| = \left[R^2 + \left(\omega L-\frac{1}{\omega C}\right)^2\right]^{1/2} = \sqrt{2}R$$

whence $\omega L - 1/\omega C = \pm R$, which has the two admissible solutions:

$$\omega_1 = \left(\frac{R^2}{4L^2}+\frac{1}{LC}\right)^{1/2} - \frac{R}{2L} \quad \text{and} \quad \omega_2 = \left(\frac{R^2}{4L^2}+\frac{1}{LC}\right)^{1/2} + \frac{R}{2L} \quad (7.9)$$

The linewidth is thus

$$\omega_2 - \omega_1 = \frac{R}{L} = \frac{\omega_0}{Q} \quad \text{or} \quad Q = \frac{\omega_0}{\omega_2-\omega_1}$$

which agrees with (7.8). However, the half-power points are symmetrical about the angular frequency $(R^2/4L^2 + 1/LC)^{1/2}$ and not about $\omega_0 = 1/\sqrt{LC}$.

The asymmetry of the current curve is shown in Fig. 7.3 where $|I|$ is plotted and the maximum value I_m decreases with increasing R for constant values of L and C. The *resonance curves* shown are also referred to as the *current response curves.*

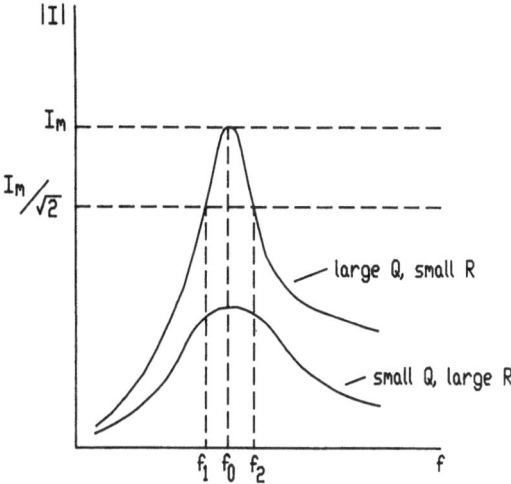

Fig. 7.3 Plots of the modulus of the current, $|I|$, as a function of frequency, f, for different magnitudes of Q.

7.2 VOLTAGE MAGNIFICATION IN A SERIES L–C–R CIRCUIT

At resonance, the potentials across the inductor and capacitor are equal, but opposite in phase, i.e.

$$V_L = j\omega_0 L I_m = \frac{j\omega_0 L V}{R} = jQV \qquad (7.10)$$

and

$$V_C = \frac{-j}{\omega_0 C} I_m = \frac{-jV}{\omega_0 CR} = -jQV \qquad (7.11)$$

Thus Q may be regarded as the factor by which the applied voltage across either L or C is magnified at resonance and the circuit behaves as a voltage transformer. Hence, Q is often referred to as the *circuit magnification factor*.

Although the maximum potential across R occurs at the resonant frequency f_0, the reader may show that the maximum potentials across L and C occur at frequencies $f_0(1 - 1/2Q^2)^{-1/2}$ and $f_0(1 - 1/2Q^2)^{1/2}$ respectively.

7.3 THE PARALLEL *L–C–R* CIRCUIT

The circuit shown in Fig. 7.4 has an admittance

$$Y = j\omega C + \frac{1}{R + j\omega L} = j\omega C + \frac{R - j\omega L}{R^2 + \omega^2 L^2} \qquad (7.12)$$

Inspection of (7.12) shows that the resonance condition, corresponding to Y real, is satisfied for $\omega = \omega_p$ where

$$R^2 + \omega_p^2 L^2 = L/C \qquad (7.13)$$

Hence

$$\omega_p = \left(\frac{1}{LC}\right)^{1/2}\left(1 - \frac{R^2 C}{L}\right)^{1/2} = \omega_0\left(1 - \frac{1}{Q^2}\right)^{1/2} \qquad (7.14)$$

where $\omega_0 = (LC)^{-1/2}$ and $Q = (1/R)\sqrt{L/C}$ are the resonant angular frequency and quality factor of the corresponding series circuit. The resonant frequency is thus

$$f_p = f_0\left(1 - \frac{1}{Q^2}\right)^{1/2} = \frac{1}{2\pi\sqrt{LC}}\left(1 - \frac{1}{Q^2}\right)^{1/2} \qquad (7.15)$$

Equations (7.12) and (7.13) give the admittance at resonance as

$$Y_p = \frac{R}{R^2 + \omega_p^2 L^2} = \frac{RC}{L} = \frac{1}{Q^2 R} \qquad (7.16)$$

corresponding to an impdance

$$Z_p = Q^2 R \qquad (7.17)$$

which is called the *parallel* or *dynamic impedance*. (If $R = \sqrt{L/C}$,

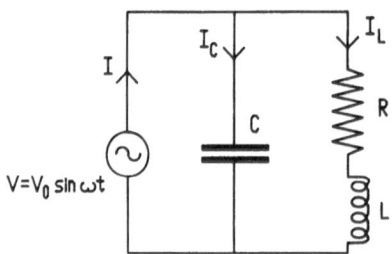

Fig. 7.4 A parallel *L–C–R* circuit.

(7.14) shows $\omega_p = 0$ and the circuit is non-resonant: this result is used in the production of feeder lines which have a magnification factor of unity and an impedance R at all frequencies.)

Equation (7.15) shows that $f_p \neq f_0$; however, in most cases, Q will be sufficiently large that the difference between f_p and f_0 can be neglected. Then

$$Q = \frac{1}{R}\sqrt{\frac{L}{C}} = \frac{\omega_0 L}{R} = \frac{1}{\omega_0 CR} \simeq \frac{\omega_p L}{R} \quad \text{or} \quad \frac{1}{\omega_p CR} \qquad (7.18)$$

It is useful to have an approximate expression for $|Y|$ near resonance, similar in form to (7.4) for the series circuit. Rearranging (7.12) gives

$$Y = \frac{R + j\omega L - 1/j\omega C}{(1/j\omega C)(R + j\omega L)} \qquad (7.19)$$

If $Q \gg 1$, $R \ll \omega L$, although $R \nleqslant \omega L - 1/\omega C$, so the term R in the denominator of (7.19) can be neglected. Then

$$Y = \frac{C}{L}\left[R + j\left(\omega L - \frac{1}{\omega C} \right) \right] \qquad (7.20)$$

which, by comparison with (7.1(a)) and (7.4) can be written as

$$Y = \frac{CR}{L}(1 + jQx) = Y_p(1 + jQx) \qquad (7.21)$$

where $x = f/f_0 - f_0/f$. Then

$$|Y| = Y_p(1 + Q^2 x^2)^{1/2} \qquad (7.22)$$

The behaviour of $|Y|$ near resonance is similar to that of the impedance for a series resonant circuit. $|Y|$ rises by a factor $\sqrt{2}$ when the frequency differs from resonance by $\pm \Delta f$ where

$$\frac{\Delta f}{f_0} = \frac{\Delta \omega}{\omega_0} = \pm \frac{1}{2Q}$$

The variation of $|Y|$ with frequency is shown in Fig. 7.5. The resonant frequency f_p decreases with decreasing Q for the same value of f_0 (7.15) and the linewidth increases.

As long as $Q \gg 1$, the current drawn from the source, and hence the power in the circuit, can be considered to be a minimum at resonance. The derivation of the exact minimum is complex and, in practical terms, unrewarding.

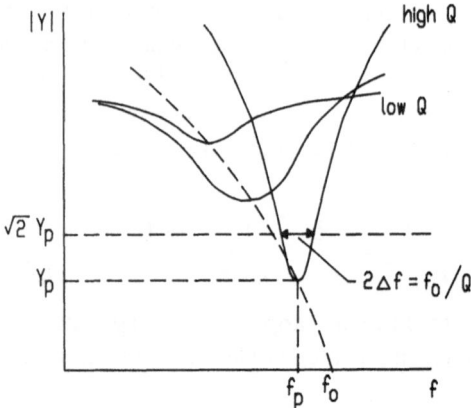

Fig. 7.5 Variation of $|Y|$ with frequency for a parallel L–C–R circuit.

7.4 CURRENT MAGNIFICATION IN A PARALLEL L–C–R CIRCUIT

If, in the circuit shown in Fig. 7.4, $I_C = j\omega_p CV$ is the current flowing through the capacitor C at resonance, and $I_p = V Y_p = V(Q^2 R)^{-1}$ is the corresponding current drawn from the source, then

$$\frac{I_C}{I_p} = \frac{j\omega_p CV}{V(Q^2R)^{-1}} = j\omega_p CRQ^2 = jQ \tag{7.23}$$

Similarly, if R is small, i.e. $Q \gg 1$, $I_L = V/j\omega_p L$ is the current flowing through the inductor L at resonance; hence

$$\frac{I_L}{I_p} = \frac{V}{j\omega_p L V(Q^2R)^{-1}} \frac{1}{} = -jQ \tag{7.24}$$

Thus the currents $|I_C|$ and $|I_L|$ are very nearly equal, are magnified by a factor Q compared with the current I_p drawn from the source and are very nearly π out of phase. The actual phase difference is more correctly approximated by $\pi - 1/Q$, where $\pi/2$ is the phase angle between $|I_C|$ and I_p and $\pi/2 - 1/Q$ is the phase angle between $|I_L|$ and I_p.

$|I_C|$ and $|I_L|$ are often referred to as the *oscillatory currents*.

7.5 ALTERNATIVE DEFINITIONS FOR RESONANCE

The definition for resonance as already used for both the series and the parallel circuits, i.e. the impedance or admittance is real, is not the only way whereby resonance may be defined in an electrical circuit. An alternative definition could be the point at which either $|Z|$, in the series circuit, or $|Y|$, in the parallel circuit, is a minimum. Obviously, the condition might also be dependent upon the component in the circuit whose value was being varied, i.e. L, C, ω or R.

For the series circuit, if L, C or ω is varied, then $|Z|$ will be a minimum when $f = f_0 = (1/2\pi)\sqrt{1/LC}$, corresponding to (7.2). However, for the parallel circuit, the condition for $|Y|$ to be a minimum is dependent upon which quantity is being varied. If f_L, f_C, f_ω and f_R are the resonant frequencies corresponding to a minimum value for $|Y|$ when L, C, ω and R respectively are varied, then it can be shown that

$$f_L = f_0 \left(1 + \frac{1}{Q^2} \right)^{1/2} \tag{7.25}$$

$$f_C = f_0 \left(1 - \frac{1}{Q^2} \right)^{1/2} = f_p \tag{7.26}$$

$$f_\omega = f_0 \left[\left(\frac{2}{Q^2} + 1 \right)^{1/2} - \frac{1}{Q^2} \right] \tag{7.27}$$

$$f_R = \frac{1}{\sqrt{2}} f_0 \tag{7.28}$$

where $Q = (1/R)\sqrt{L/C}$. Only when C is varied is the resonant frequency, f_C, equal to that of the current with a power factor of unity, i.e. $f_C = f_p$. When either L or ω is varied, slightly different resonant conditions are obtained, but in all three cases the resonant frequency tends to f_0 when $Q \gg 1$. Only when R is varied is the resonant frequency, f_R, independent of Q.

It is left as an exercise for the reader to show that the conditions represented by (7.25)–(7.28) are correct.

7.6 USES OF SERIES AND PARALLEL RESONANT CIRCUITS

The sharp response of the series circuit, with a minimum impedance at resonance, enables it to be used to select a narrow band of frequencies from a wide range and hence the series circuit is often referred to as an *accepter circuit*. In contrast, the sharp response of the parallel circuit, with a high impedance at resonance, enables it to be used to reject a narrow band of frequencies and consequently it is often called a *rejector circuit*. Thus the series and parallel resonant circuits are complementary and can be used as band-pass and band-stop filters respectively. The properties of the parallel tuned circuit, i.e. a sharp maximum impedance at resonance, are often utilized in narrow-band amplifiers, thus enabling high amplification to be achieved over a narrow band of frequencies.

7.7 DEFINITIONS OF Q

From consideration of the series and parallel resonant circuits it will be seen that the quality or magnification factor, Q, can be defined in a variety of ways.

1. From (7.18),

$$Q = \frac{1}{R}\sqrt{\frac{L}{C}} = \frac{\omega_0 L}{R} = \frac{1}{\omega_0 CR} \simeq \frac{\omega_p L}{R} \quad \text{or} \quad \frac{1}{\omega_p CR}$$

2. Q is the voltage magnification obtained by using the series circuit as a tuned transformer (cf. (7.10) and (7.11)).
3. Q is the ratio of the oscillatory current to the supply current in the parallel resonant circuit (cf. (7.23) and (7.24)).
4. The parallel or dynamic resistance, Z_p of a circuit is Q^2 times its series resistance (cf. (7.17)).
5. The fractional frequency difference, $2\Delta f/f_0$, between the points at which the impedance or admittance changes by a factor $\sqrt{2}$ from the value at resonance is equal to the reciprocal of Q (cf. (7.8)).
6. The same definition as 5 applies if the power dissipated changes by a factor 2: hence the reference to the linewidth $2\Delta f$ occurring at the half-power points.

In more complicated circuits, and in resonant waveguide cavities where values for L, C and R cannot be specified, it is useful to define the Q factor in terms of energy storage and dissipation; thus

$$Q = \frac{\omega \times \text{stored energy in circuit}}{\text{the average rate of energy dissipation}}$$
$$= \frac{\text{stored energy in circuit}}{\text{energy dissipated per radian}} \tag{7.29}$$

It can be readily shown that this definition agrees with that of (7.18) for a series resonant circuit. Thus, at resonance, the energy stored at any instant, when a current I flows through the circuit and the capacitor has a charge q, is

$$u = \tfrac{1}{2}LI^2 + \tfrac{1}{2}\frac{q^2}{C} = \tfrac{1}{2}LI^2 \sin^2 \omega_0 t + \tfrac{1}{2}\frac{I_0^2}{\omega_0 C} \cos^2 \omega_0 t$$

But $\omega_0^2 = 1/LC$, so that

$$u = \tfrac{1}{2}LI_0^2$$

The average rate at which energy is dissipated in the circuit is $\tfrac{1}{2}I_0^2 R$; hence, from (7.29),

$$Q = \frac{\omega_0 \times \tfrac{1}{2}LI_0^2}{\tfrac{1}{2}I_0^2 R} = \frac{\omega_0 L}{R}$$

8

Coupled circuits

Two circuits are said to be coupled if they are linked by a common impedance such as a resistance, a capacitance, a self-inductance or a mutual inductance. In this chapter, a generalized coupled circuit is considered first, and the results are then extended to the particular cases of (i) a low frequency transformer, (ii) resonant circuits coupled by a mutual inductance and (iii) resonant circuits directly coupled by a capacitance.

8.1 THE GENERALIZED COUPLED CIRCUIT

Consider the circuit shown in Fig. 8.1 in which a potential V_1 is applied to the terminals 11' of the primary circuit and a load of impedance Z_L is connected to the terminals 22' of the secondary circuit. The coupling impedance between the two circuits is taken to be Z_c. The total primary impedance is

$$Z_{11} = Z_1 + Z_c$$

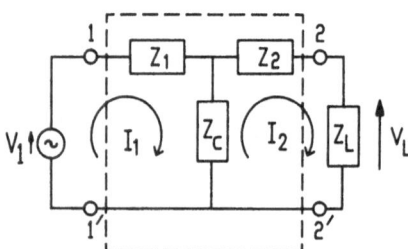

Fig. 8.1 General representation of a coupled circuit.

and the total secondary impedance is

$$Z_{22} = Z_2 + Z_c + Z_L$$

If I_1 and I_2 are the currents in the primary and secondary circuits respectively, then the matrix representing the circuit conditions is (see (5.8))

$$\begin{bmatrix} Z_{11} & -Z_c \\ -Z_c & Z_{22} \end{bmatrix} \begin{bmatrix} I_1 \\ I_2 \end{bmatrix} = \begin{bmatrix} V_1 \\ 0 \end{bmatrix}$$

Hence

$$I_1 = \frac{V_1 Z_{22}}{Z_{11}Z_{22} - Z_c^2} \tag{8.1}$$

and

$$I_2 = \frac{V_1 Z_c}{Z_{11}Z_{22} - Z_c^2} \tag{8.2}$$

The potential which appears across the load is then

$$V_L = I_2 Z_L = \frac{V_1 Z_c Z_L}{Z_{11}Z_{22} - Z_c^2} \tag{8.3}$$

Examination of (8.1) and (8.2) shows that the generalized coupled circuit shown in Fig. 8.1 may be represented by two separate equivalent circuits as shown in Fig. 8.2. In these two circuits, $-Z_c^2/Z_{22}$ represents the impedance 'reflected' into the primary circuit from the secondary circuit, whilst $-Z_c^2/Z_{11}$ represents the impedance 'reflected' into the secondary circuit from the primary circuit.

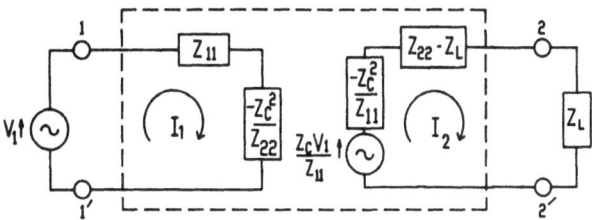

Fig. 8.2 Equivalent circuits corresonding to the primary and secondary circuits of the generalized coupled circuit of Fig. 8.1.

8.2 THE LOW FREQUENCY TRANSFORMER

A practical low frequency transformer consists of a primary coil and a secondary coil which are closely wound together on an iron core so that almost all of the magnetic flux due to a current in one coil will pass through the other. At low frequencies the effects of stray capacitances are small and the transformer provides an efficient method for transferring power from one circuit to another. Such a transformer is represented by the circuit shown in Fig. 8.3(a). The reactances of the coils are often much greater than any other impedances in the primary and secondary circuits; hence it will be assumed that $Z_1 \ll \omega L_1$ and $Z_2 \ll \omega L_2$.

The equivalent circuit (see section 6.9) of the low frequency transformer is shown in Fig. 8.3(b) and, by comparing this circuit with that shown in Fig. 8.1, it will be seen that

$$Z_{11} = Z_1 + j\omega L_1, \quad Z_{22} = Z_2 + j\omega L_2$$
$$Z_c = -j\omega M, \qquad Z_L = 0 \tag{8.4}$$

(Note $Z_L = 0$ because it has now been incorporated in Z_2.) Substituting in (8.1) and (8.2) and rearranging gives

$$\frac{V_1}{I_1} = Z_{11} - \frac{Z_c^2}{Z_{22}} = Z_1 + j\omega L_1 + \frac{\omega^2 M^2}{Z_2 + j\omega L_2} \tag{8.5}$$

and

$$\frac{Z_c V_1}{I_2} = Z_{11} Z_{22} - Z_c^2$$

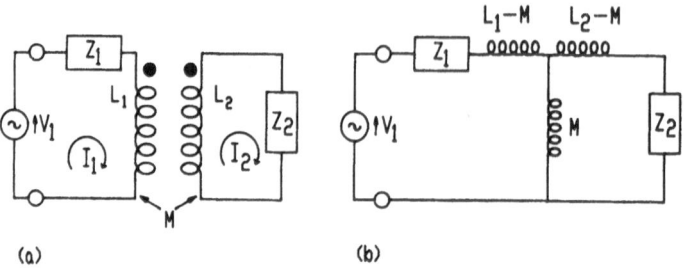

(a)   (b)

Fig. 8.3 (a) The low frequency transformer, and (b) the equivalent circuit.

so that

$$\frac{-j\omega M V_1}{I_2} = (Z_1 + j\omega L_1)(Z_2 + j\omega L_2) + \omega^2 M^2$$

Dividing by $j\omega L_1$ gives

$$\frac{-(M/L_1)V_1}{I_2} = \frac{Z_1 Z_2}{j\omega L_1} + Z_2 + Z_1 \frac{L_2}{L_1} + j\omega\left(L_2 - \frac{M^2}{L_1}\right)$$

$$\simeq Z_2 + Z_1 \frac{L_2}{L_1} + j\omega\left(L_2 - \frac{M^2}{L_1}\right) \tag{8.6}$$

Since $Z_1 Z_2 / j\omega L_1 \ll Z_1$.

The self-inductance of a coil is proportional to the square of the number of turns; hence the ratio of the number of turns on the secondary coil to the number of turns on the primary coil can be written as $n = (L_2/L_1)^{1/2}$. Also, for a *perfect transformer* with complete coupling (1.1) gives $M = (L_1 L_2)^{1/2}$, so that

$$\frac{M}{L_1} = \frac{L_2}{M} = \left(\frac{L_2}{L_1}\right)^{1/2} = n$$

It follows from (8.5), and remembering also that $Z_2 \ll \omega L_2$, that the impedance of the primary circuit can be written as

$$Z_p = \frac{V_1}{I_1} = Z_1 + j\omega L_1 + \frac{\omega^2 M^2 (Z_2 - j\omega L_2)}{Z_2^2 + \omega^2 L_2^2}$$

$$\simeq Z_1 + Z_2\left(\frac{M}{L_2}\right)^2 + j\omega\left(L_1 - \frac{M^2}{L_2}\right)$$

which, since $M^2 = L_1 L_2$, gives a primary impedance

$$Z_p = Z_1 + \frac{Z_2}{n^2} \tag{8.7}$$

From (8.6), the current in the secondary circuit can be considered to be due to an emf of $(-M/L_1)V_1 = -nV_1$ connected to the secondary impedance Z_s where

$$Z_s = Z_2 + Z_1 n^2 \tag{8.8}$$

Equations (8.7) and (8.8), which have been derived for a perfect transformer on load such that $Z_2 \ll \omega L_2$, show that the effect of the secondary circuit on the primary current is represented by the

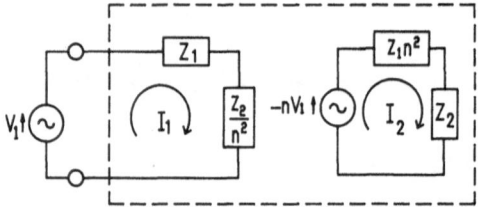

Fig. 8.4 The equivalent circuits corresponding to the primary and secondary circuits of a perfect transformer.

additional *reflected impedance* Z_2/n^2, whilst, in the secondary circuit, there is an effective source of emf equal to $-nV_1$ with an apparent internal impedance of $Z_1 n^2$. The equivalent circuits corresponding to the primary and secondary circuits of a perfect transformer are shown in Fig. 8.4.

Finally, for a perfect transformer, it can be seen that

$$I_1 = \frac{V_1}{Z_p} = \frac{V_1}{Z_1 + Z_2/n^2}$$

and

$$I_2 = \frac{-nV_1}{Z_s} = \frac{-nV_1}{Z_2 + Z_1 n^2} = \frac{-V_1/n}{Z_1 + Z_2/n^2} = -\frac{1}{n}I_1$$

Hence the currents are transformed by $1/n$, a result which should be expected for a lossless transformer since the power in the two circuits must be the same, i.e. $V_1 I_1 = V_2 I_2$.

In practice not all of the flux from one circuit passes through the other, so that M^2 is slightly less than $L_1 L_2$ and

$$M = k(L_1 L_2)^{1/2}$$

where k is the coefficient of coupling and has a value close to unity. Equation (8.5) can then be rewritten as

$$Z_p = \frac{V_1}{I_1} = Z_1 + j\omega(1-k)L_1 + \frac{\omega^2 k^2 L_1 L_2 + j\omega L_1 k(Z_2 + j\omega L_2)}{Z_2 + j\omega L_2}$$

$$= Z_1 + j\omega(1-k)L_1 + \frac{j\omega k L_1 + [Z_2 + j\omega(1-k)L_2]}{Z_2 + j\omega(1-k)L_2 + j\omega k L_1 n^2}$$

$$= Z_1 + j\omega(1-k)L_1 + \left[\frac{1}{j\omega k L_1} + \frac{n^2}{Z_2 + j\omega(1-k)L_2}\right]^{-1} \quad (8.9)$$

where use has been made of the identity $n^2 L_1 = L_2$.

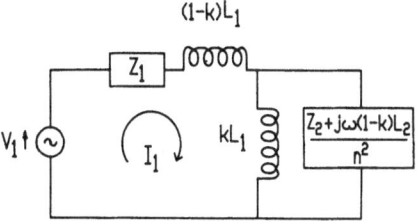

Fig. 8.5 Representation of the primary circuit of a practical low frequency transformer.

The last term in (8.9) represents two impedances in parallel and, consequently, the primary circuit can be represented by the circuit shown in Fig. 8.5. In this circuit, $(1 - k)L_1$ represents the leakage inductance due to imperfect coupling and the impedance

$$\frac{Z_2 + j\omega(1 - k)L_2}{n^2}$$

is reflected from the secondary circuit in parallel with the remainder of the primary inductance, i.e. kL_1. Assuming that $Z_1 \ll j\omega L_1$ and substituting for $M/L_1 = k(L_2/L_1)^{1/2} = kn$ in (8.6) and rearranging, gives

$$\frac{-knV_1}{I_2} = Z_2 + n^2 Z_1 + j\omega(1 - k)L_2 + j\omega\left(kL_2 - \frac{M^2}{L_1}\right)$$

$$= Z_2 + n^2[Z_1 + j\omega(1 - k)L_1] + j\omega k(1 - k)L_2 \quad (8.10)$$

In (8.10) the term $n^2[Z_1 + j\omega(1 - k)L_1]$ represents the impedance reflected from the primary circuit and $k(1 - k)L_2$ is the secondary leakage inductance. Thus the secondary circuit can be represented

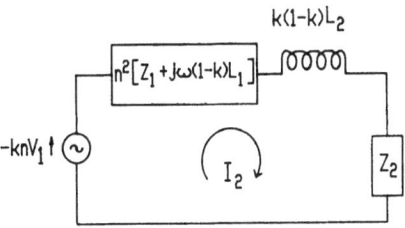

Fig. 8.6 Circuit representing the secondary circuit of a practical low frequency transformer.

as shown in Fig. 8.6. It should be noted that the secondary leakage inductance differs, by a factor k, from that which appears in series with Z_2 in the expression for the impedance reflected from the secondary circuit into the primary circuit. However, the two expressions for the secondary leakage inductance are approximately equal if $k \to 1$ and since, in most cases, the secondary leakage inductance will be small compared with Z_2, it is usually approximated to $(1 - k)L_2$.

The complete equivalent circuit of the transformer is shown in Fig. 8.7. The portion of the circuit enclosed in the broken rectangle represents a perfect transformer. In addition Z_2 has been replaced by Z_L and r_2 in series, where r_2 is the resistance of the secondary coil and Z_L is the impedance of the load connected to the terminals of the secondary circuit. Also r_1 represents the resistance of the primary coil which previously has been regarded as part of Z_1. r_1 and r_2 allow for the dissipation of energy in the coils of the transformer, i.e. the *copper losses*, whilst the resistance R in parallel with kL allows for the loss of energy through hysteresis and eddy currents in the core, i.e. the *iron losses*.

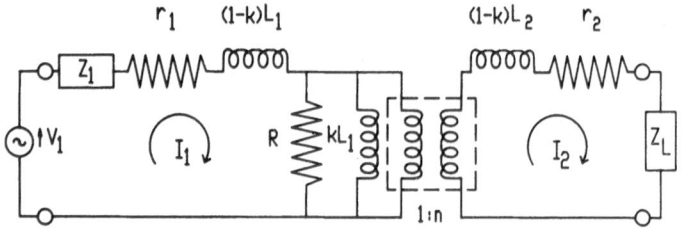

Fig. 8.7 Complete equivalent circuit for a practical low frequency transformer.

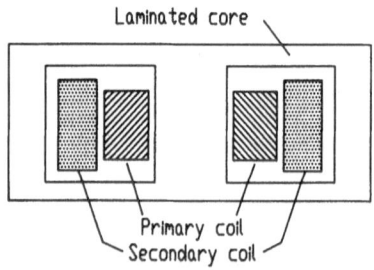

Fig. 8.8 Construction of a low frequency transformer.

The physical construction of a low frequency transformer is shown in Fig. 8.8. The coils are wound on a ferromagnetic core which reduces the magnetic leakage, i.e. $k \to 1$. The iron losses are reduced by using a material for the core which has a narrow hysteresis loop, e.g. silicon steel, and by making the core from a series of thin laminations which are insulated from each other. The insulation increases the resistance of the paths taken by the eddy currents, thus reducing their magnitude.

8.3 RESONANT CIRCUITS COUPLED BY A MUTUAL INDUCTANCE

Consider two $L-C-R$ circuits which have the same resonant frequency and which are coupled together through a mutual inductance as shown in Fig. 8.9(a). The equivalent circuit of this system (see section 6.9) is shown in Fig. 8.9(b).

If the equivalent circuit is compared with the general form shown in Fig. 8.1 it will be seen that, at an angular frequency ω, the impedances in the latter can be written explicitly as

$$Z_{11} = R_1 + j\left(\omega L_1 - \frac{1}{\omega C_1}\right) = R_1 + jX_1$$

$$Z_{22} = R_2 + j\left(\omega L_2 - \frac{1}{\omega C_2}\right) = R_2 + jX_2 \qquad (8.11)$$

$$Z_c = -j\omega M, \quad Z_L = 0$$

and the resonant conditions for the circuits are that, at ω_0, $X_1 = 0$ and $X_2 = 0$, or $L_1 C_1 = L_2 C_2 = 1/\omega_0^2$.

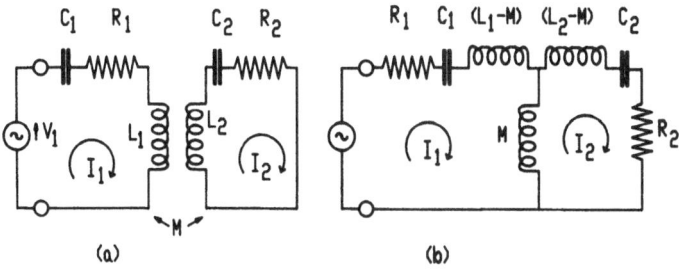

(a)   (b)

Fig. 8.9 (a) Resonant circuits coupled by a mutual inductance. (b) The equivalent circuit.

It follows from (8.1) that

$$I_1 = \frac{V_1 Z_{22}}{Z_{11}Z_{22} - Z_c^2}$$

$$= \frac{V_1(R_2 + jX_2)}{(R_1 + jX_1)(R_2 + jX_2) + \omega^2 M^2} \tag{8.12}$$

$$= \frac{V_1}{R_p + jX_p} \tag{8.13}$$

where

$$R_p = \frac{R_1(R_2^2 + X_2^2) + R_2\omega^2 M^2}{R_2^2 + X_2^2}$$

and

$$X_p = \frac{X_1(R_2^2 + X_2^2) - X_2\omega^2 M^2}{R_2^2 + X_2^2}$$

Similarly, from (8.2)

$$I_2 = \frac{-j\omega M V_1}{(R_1 + jX_1)(R_2 + jX_2) + \omega^2 M^2} \tag{8.14}$$

From (8.13) it will be seen that V_1 and I_1 are in phase if $X_p = 0$, i.e.

$$X_1(R_2^2 + X_2^2) - X_2\omega^2 M^2 = 0 \tag{8.15}$$

It should be noted that a particular solution of (8.15) occurs when $\omega = \omega_0$, i.e. when $X_1 = X_2 = 0$.

Consider now the currents that will flow in the two circuits (1) at resonance ($\omega = \omega_0$), (2) near resonance ($\omega \simeq \omega_0$) and (3) far from resonance ($\omega \not\simeq \omega_0$).

8.3.1 Currents at resonance

From (8.12) and (8.14), when $\omega = \omega_0$,

$$I_1 = \frac{R_2 V_1}{R_1 R_2 + \omega_0^2 M^2} = |I_1| \tag{8.16}$$

$$I_2 = \frac{-j\omega_0 M V_1}{R_1 R_2 + \omega_0^2 M^2} = -j|I_2| \tag{8.17}$$

It should be noted that I_2 lags I_1 by a phase angle of $\pi/2$.

If M is steadily increased from zero by increasing the coupling coefficient k, then the primary current $|I_1|$ decreases steadily from its original value of V_1/R_1, whereas the secondary current $|I_2|$ increases from zero, passes through a maximum and then decreases. The maximum is given by the condition

$$\frac{d|I_2|}{dt} = \frac{\omega_0 V_1[R_1R_2 + \omega_0^2 M^2 - M(2\omega_0^2 M)]}{(R_1R_2 + \omega_0^2 M^2)^2} = 0$$

or

$$\omega_0 M = (R_1 R_2)^{1/2} \tag{8.18}$$

Equation (8.18) represents the condition for *optimum coupling* and if k_0 is the corresponding coupling coefficient then,

$$k_0 = \frac{M}{(L_1 L_2)^{1/2}} = \frac{1}{\omega_0}\left(\frac{R_1 R_2}{L_1 L_2}\right)^{1/2} = \frac{1}{(Q_1 Q_2)^{1/2}} \tag{8.19}$$

where Q_1 and Q_2 are the Q factors of the corresponding circuits.

At optimum coupling, the impedance reflected from the secondary circuit into the primary circuit is, from Fig. 8.2,

$$\frac{-Z_c^2}{Z_{22}} = \frac{\omega_0^2 M^2}{R_2} = \frac{R_1 R_2}{R_2} = R_1$$

and the secondary circuit is matched to the primary circuit.

Also from (8.16) and (8.17), at optimum coupling,

$$|I_1| = \frac{R_2 V_1}{R_1 R_2 + \omega_0^2 M^2} = \frac{V_1}{2R_1} \tag{8.20}$$

and

$$|I_2| = \frac{\omega_0 M V_1}{R_1 R_2 + \omega_0^2 M^2} = \frac{V_1}{2(R_1 R_2)^{1/2}} \tag{8.21}$$

8.3.2 Currents near resonance

Suppose that $\omega = \omega_0 + \Delta\omega$ where $|\Delta\omega| \ll \omega_0$: then

$$X_1 = (\omega_0 + \Delta\omega)L_1 - \frac{1}{(\omega_0 + \Delta\omega)C_1}$$

$$\simeq (\omega_0 + \Delta\omega)L_1 - \frac{1}{\omega_0 C_1}\left(1 - \frac{\Delta\omega}{\omega_0}\right)$$

or

$$X_1 \simeq 2\,\Delta\omega\,L_1 \tag{8.22}$$

Similarly

$$X_2 \simeq 2\,\Delta\omega\,L_2 \tag{8.23}$$

8.3.2.1 *The conditions for the secondary current to be either a maximum or a minimum*

In order to find the conditions for the secondary current to be either a maximum or a minimum, it is necessary first to obtain the magnitude of the secondary current by rationalizing the expression given by (8.14); then

$$|I_2| = \frac{\omega M V_1}{[(R_1 R_2 - X_1 X_2 + \omega^2 M^2)^2 + (X_1 R_2 + X_2 R_1)^2]^{1/2}} \tag{8.24}$$

and, in the vicinity of resonance where (8.22) and (8.23) hold,

$$|I_2| = \frac{\omega_0 M V_1}{\{[R_1 R_2 - 4L_1 L_2 (\Delta\omega)^2 + \omega_0^2 M^2]^2 + 4(\Delta\omega)^2 (L_1 R_2 + L_2 R_1)^2\}^{1/2}} \tag{8.25}$$

Since $\Delta\omega$ only appears in the denominator of (8.25) then $|I_2|$ will be a maximum or a minimum when

$$\frac{d(\text{denominator})}{d(\Delta\omega)} = 0$$

which, on performing the differentiation and equating to zero, gives

$$8L_1^2 L_2^2 (\Delta\omega)^3 + (L_1^2 R_2^2 + L_2^2 R_1^2 - 2L_1 L_2 \omega_0^2 M^2)\,\Delta\omega = 0$$

which is satisfied by either

$$\Delta\omega = 0$$

or

$$4(\Delta\omega)^2 = \frac{\omega_0^2 M^2}{L_1 L_2} - \tfrac{1}{2}\left(\frac{R_1^2}{L_1^2} + \frac{R_2^2}{L_2^2}\right)$$

The second condition gives

$$\Delta\omega = \pm\frac{\omega_0}{2}\left[k^2 - \tfrac{1}{2}\left(\frac{1}{Q_1^2} + \frac{1}{Q_2^2}\right)\right]^{1/2} \tag{8.26}$$

If $k^2 > \frac{1}{2}(1/Q_1^2 + 1/Q_2^2)$ there are three real roots, $\Delta\omega = 0$ being a minimum and the other two, given by (8.26) being maxima; however, if $k^2 < \frac{1}{2}(1/Q_1^2 + 1/Q_2^2)$ there is only the real root corresponding to $\Delta\omega = 0$, which is a maximum.

Critical coupling occurs when all three roots are zero; hence the *critical coupling coefficient*, k_C, is given by

$$k_c^2 = \frac{1}{2}\left(\frac{1}{Q_1^2} + \frac{1}{Q_2^2}\right) \tag{8.27}$$

and $\Delta\omega = 0$ is a point of inflexion. Note that, from comparison with (8.19), $k_c \gg k_0$ since

$$2(k_c^2 - k_0^2) = \frac{1}{Q_1^2} + \frac{1}{Q_2^2} - \frac{2}{Q_1 Q_2} = \left(\frac{1}{Q_1} - \frac{1}{Q_2}\right)^2 \geqslant 0$$

8.3.2.2 *The condition for the primary impedance to be real*

The general condition for the primary impedance to be real is given by (8.15) and, in the vicinity of resonance ($\omega \simeq \omega_0$) where (8.22) and (8.23) hold, this equation approximates to

$$2\,\Delta\omega\,L_1 R_2^2 + 2\,\Delta\omega\,L_1 \times 4(\Delta\omega)^2 L_2^2 - 2\,\Delta\omega\,L_2\omega_0^2 M^2 = 0$$

which is satisfied by $\Delta\omega = 0$ or $4(\Delta\omega)^2 L_1 L_2^2 = L_2\omega_0^2 M^2 - L_1 R_2^2$. The second of these conditions gives

$$\Delta\omega = \pm\frac{\omega_0}{2}\left(k^2 - \frac{1}{Q_2^2}\right)^{1/2} \tag{8.28}$$

8.3.2.3 *The condition for the secondary impedance to be real*

By rearranging (8.14) the secondary current can be expressed as

$$I_2 = \frac{-\omega M V_1}{X_1 R_2 + X_2 R_1 - j(R_1 R_2 - X_1 X_2 + \omega^2 M^2)}$$

and the secondary current will be real when

$$R_1 R_2 - X_1 X_2 + \omega^2 M^2 = 0 \tag{8.29}$$

In the vicinity of resonance, where (8.22) and (8.23) hold, this equation approximates to

$$R_1 R_2 - 4(\Delta\omega)^2 L_1 L_2 + \omega_0^2 M^2 = 0 \tag{8.30}$$

and

$$\Delta\omega = \pm \frac{\omega_0}{2}\left(\frac{M^2}{L_1 L_2} + \frac{R_1 R_2}{\omega_0^2 L_1 L_2}\right)^{1/2} = \pm \frac{\omega_0}{2}(k^2 + k_0^2)^{1/2} \quad (8.31)$$

where $k_0^2 = (Q_1 Q_2)^{-1}$ from (8.19).

For circuits of reasonably high Q, the conditions that $|I_2|$ is a maximum (given by (8.26)), that the primary impedance is real (given by (8.28)), and that the secondary impedance is real (given by (8.31)), all approximate to the same value given by

$$\Delta\omega = \pm \frac{\omega_0 M}{2(L_1 L_2)^{1/2}} \quad (8.32)$$

Therefore, for high Q circuits in the vicinity of resonance, the two maxima for the secondary current occur at frequencies where both the primary and secondary impedances may be considered, to a good approximation, to be real. At both maxima, the secondary current will have approximately the same value, which can be obtained by substituting from (8.30) and (8.32) into (8.25), whence

$$|I_2|_{\text{max}} = \frac{(L_1 L_2)^{1/2} V_1}{(L_1 R_2 + L_2 R_1)^{1/2}} \quad (8.33)$$

Corresponding conditions can be obtained for the primary current $|I_1|$ and it will be found that $|I_1|$ behaves in a similar manner to $|I_2|$.

8.3.3 Current at frequencies remote from ω_0

It is clear from (8.32) that, as the coupling between the circuits increases, so the separation of the peaks in $|I_2|$ will increase and the condition $\Delta\omega \ll \omega_0$ will fail. However, in this case, the resistances of the circuits will generally become negligible compared with the reactances and (8.12) and (8.14) become

$$I_1 \simeq \frac{jX_2 V_1}{j(R_1 X_2 + R_2 X_1) + \omega^2 M^2 - X_1 X_2}$$

and

$$I_2 \simeq \frac{-j\omega M V_1}{j(R_1 X_2 + R_2 X_1) + \omega^2 M^2 - X_1 X_2} \quad (8.34)$$

The maxima in $|I_1|$ and $|I_2|$ occur very nearly when $\omega^2 M^2 = X_1 X_2$, i.e. when

$$\omega^2 k^2 L_1 L_2 - \left(\omega L_1 - \frac{1}{\omega C_1}\right)\left(\omega L_2 - \frac{1}{\omega C_2}\right) = 0$$

which, on putting $1/C_1 = \omega_0^2 L_1$ and $1/C_2 = \omega_0^2 L_2$, gives

$$\omega = \frac{\omega_0}{(1 \pm k)^{1/2}} \qquad (8.35)$$

Equation (8.35) gives the positions of the two peaks when the coupling is tight. From (8.34), it will be seen that, at the peaks, I_1 is in phase with V_1 whilst I_2 is in antiphase with V_2.

The behaviour of the primary and secondary currents in a circuit having mutual inductance coupling is shown in Fig. 8.10. All types of coupled circuits behave in a broadly similar manner, i.e. loose coupling means that the circuits are virtually independent whereas tighter coupling produces a double resonance in both circuits at frequencies which become more widely separated as the coupling increases.

8.4 RESONANT CIRCUITS WITH DIRECT COUPLING

Circuits can also be coupled directly by means of a self-inductance or a capacitance. Direct coupling via a capacitance C_c is shown in

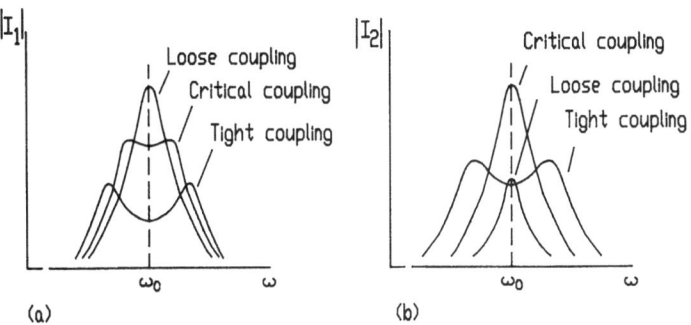

Fig. 8.10 (a) Variation of the primary current when two resonant circuits are coupled by a mutual inductance. (b) Corresponding variation of the secondary current.

Fig. 8.11, and, in this case,

$$C_1' = \frac{C_1 C_c}{C_1 + C_c} \quad \text{and} \quad C_2' = \frac{C_2 C_c}{C_2 + C_c}$$

where C_1' and C_2' are the total capacitances in the primary and secondary circuits respectively. The resonances coincide for $\omega_0^2 = 1/L_1 C_1' = 1/L_2 C_2'$. In such a circuit it is necessary to use a more generalized definition for the coupling coefficient, k, namely

$$k = \frac{X_c}{(X_1 X_2)^{1/2}} \tag{8.36}$$

where X_c is the coupling reactance between the circuits and X_1 and X_2 are the total reactances, of the same kind, in the primary and secondary circuits. Hence, for the circuit shown in Fig. 8.10

$$k = \frac{(C_1' C_2')^{1/2}}{C_c} \tag{8.37}$$

For frequencies remote from ω_0, the resistances in the two circuits can be regarded as being negligible compared with the reactances; therefore, from (8.1) and (8.2), the currents in the two circuits will be maxima when

$$Z_{11} Z_{22} - Z_c^2 = 0$$

or

$$j^2 \left(\omega L_1 - \frac{1}{\omega C_1'} \right) \left(\omega L_2 - \frac{1}{\omega C_2'} \right) - \left(\frac{1}{j\omega C_c} \right)^2 = 0$$

Substituting for L_1 and L_2 in terms of ω_0, C_1' and C_2' gives

$$\frac{1}{C_1' C_2'} \left(\frac{\omega^2 - \omega_0^2}{\omega_0^2 \omega} \right)^2 - \frac{1}{\omega^2 C_c^2} - 0$$

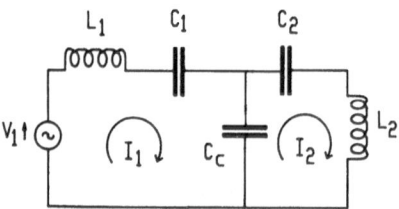

Fig. 8.11 Resonant circuits coupled directly via a capacitance C_c.

or

$$\frac{\omega^2 - \omega_0^2}{\omega_0^2} = \pm \frac{(C_1' C_2')^{1/2}}{C_c} = \pm k$$

or

$$\omega = \omega_0 (1 \pm k)^{1/2} \tag{8.38}$$

From (8.38) it can be seen that, as in the case of mutual inductance coupling, the currents will exhibit double peaks with separation proportional to the coupling coefficient.

8.5 USES OF COUPLED RESONANT CIRCUITS

Three examples of typical applications of coupled resonant circuits are described in this section.

In radio receivers, circuits coupled by a mutual inductance are used as band-pass filters. Such circuits, employing rather more than critical coupling, produce a broad response curve which falls away more rapidly at the edges than the response curve of a single resonant circuit. The response curve therefore approximates to the rectangular shape of an ideal filter and the slight central dip in the curve is compensated by the use of undercoupled circuits with a single peak elsewhere in the circuit.

It is obviously necessary for radio receivers to be tunable, and desirable for the width of the resonance to be independent of the central frequency of the resonance. This cannot be achieved by the use of mutual inductive coupling alone (see (8.26)), but it is possible to achieve a more or less constant bandwidth as the receiver is tuned by using a combination of capacitive and mutual inductive coupling.

Finally, it is possible to use the change which occurs in the resonant frequency with the variation of the coupling to tune a circuit without altering the resonating components: this can be particularly useful where the resonance occurs in a fixed resonator such as a dielectric cavity.

9

Two-port networks

A common practical situation occurs with a network when an input or a generator is applied across one branch of the network and the output or response across another branch has to be determined. In such a case, the network may be effectively represented by a 'black box' which has two input terminals (the *input port*) and two output terminals (the *output port*): this situation is illustrated in Figs 9.1(a) and 9.1(b) where, in the latter, one terminal is common, giving a three-terminal rather than a four-terminal network.

The present chapter is only concerned with passive networks which consist of combinations of resistances, capacitances and self- and mutualinductances, and in which there are no active elements generating current or voltage.

For a passive network, the relation to each other of I_1, I_2, V_1 and V_2 in Fig. 9.1 is such that, if any of these quantities are given, the other two are determined. This fact, which follows from Norton's (or Thévenin's) theorem, is illustrated in section 9.1.1, in which I_1 and I_2 are considered to be given (or are taken as independent variables) while V_1 and V_2 are derived (or dependent) variables, each

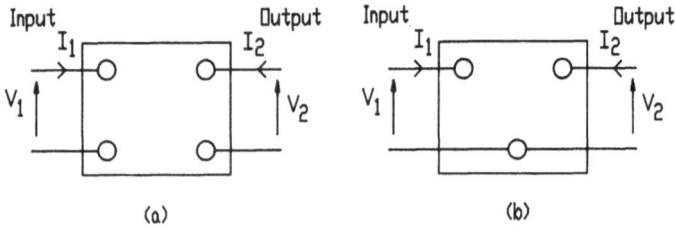

Fig. 9.1 (a) Four-terminal and (b) three-terminal two-port networks.

being some linear combination of I_1 and I_2. (There are actually six ways of choosing the pair of independent variables but only four of these ways—those which have practical application—will be considered in this chapter.)

9.1 PARAMETRIC REPRESENTATIONS OF A TWO-PORT NETWORK

9.1.1 Impedance or z parameters

With I_1 and I_2 given, it is possible to write

$$\begin{aligned} V_1 &= z_{11}I_1 + z_{12}I_2 \\ V_2 &= z_{21}I_1 + z_{22}I_2 \end{aligned} \tag{9.1}$$

where the parameters z_{ij} are constant coefficients which obviously have the dimensions of impedance. Equation (9.1) may be expressed in matrix form

$$\begin{bmatrix} V_1 \\ V_2 \end{bmatrix} = \begin{bmatrix} z_{11} & z_{12} \\ z_{21} & z_{22} \end{bmatrix} \begin{bmatrix} I_1 \\ I_2 \end{bmatrix} = [z] \begin{bmatrix} I_1 \\ I_2 \end{bmatrix} \tag{9.2}$$

where

$$[z] = \begin{bmatrix} z_{11} & z_{12} \\ z_{21} & z_{22} \end{bmatrix}$$

The four z parameters can be defined by setting either $I_1 = 0$ or $I_2 = 0$, and identified by the function which they perform:

$$z_{11} = \left(\frac{V_1}{I_1} \right)_{I_2=0} \qquad \text{open-circuit input impedance}$$

$$z_{21} = \left(\frac{V_2}{I_1} \right)_{I_2=0} \qquad \text{open-circuit forward transfer impedance}$$

$$z_{12} = \left(\frac{V_1}{I_2} \right)_{I_1=0} \qquad \text{open-circuit reverse transfer impedance} \tag{9.3}$$

$$z_{22} = \left(\frac{V_2}{I_2} \right)_{I_1=0} \qquad \text{open-circuit output impedance}$$

Equation (9.1) can be represented by the equivalent circuit of Fig. 9.2, the terms, $z_{12}I_2$ and $z_{21}I_1$ being included as 'current-dependent' voltage generators.

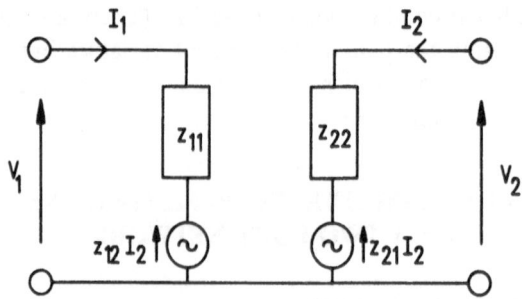

Fig. 9.2 Equivalent circuit for z parameters.

9.1.2 Admittance or y parameters

If, instead of I_1 and I_2 (as used above), it is V_1 and V_2 which are taken as given (or independent), then it is possible to write

$$I_1 = y_{11}V_1 + y_{12}V_2$$
$$I_2 = y_{21}V_1 + y_{22}V_2 \tag{9.4}$$

so that

$$\begin{bmatrix} I_1 \\ I_2 \end{bmatrix} = [y] \begin{bmatrix} V_1 \\ V_2 \end{bmatrix} \quad \text{where } [y] = \begin{bmatrix} y_{11} & y_{12} \\ y_{21} & y_{22} \end{bmatrix} \tag{9.5}$$

The four y parameters have dimensions of admittance and can be defined by putting $V_1 = 0$ or $V_2 = 0$ so that

$$y_{11} = \left(\frac{I_1}{V_1} \right)_{V_2=0} \quad \text{short-circuit input admittance} \tag{9.6}$$

and so on by analogy with (9.3).

Equation (9.4) can be represented by the equivalent circuit of Fig. 9.3.

9.1.3 Hybrid or h parameters

A very important set of parameters, the h parameters, arise if I_1 and V_2 are taken as independent. Then

$$V_1 = h_{11}I_1 + h_{12}V_2$$
$$I_2 = h_{21}I_1 + h_{22}V_2 \tag{9.7}$$

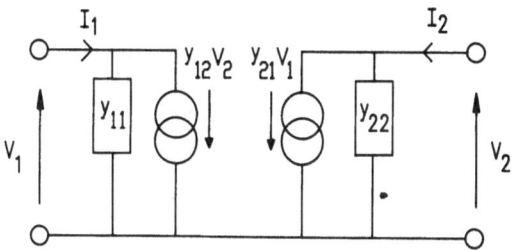

Fig. 9.3 Equivalent circuit for *y* parameters.

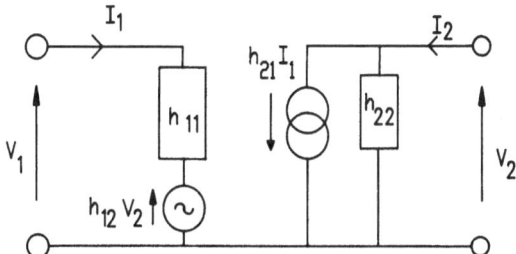

Fig. 9.4 Equivalent circuit for *h* parameters.

so that

$$\begin{bmatrix} V_1 \\ I_2 \end{bmatrix} = [h]\begin{bmatrix} I_1 \\ V_2 \end{bmatrix} \quad \text{where} \quad [h] = \begin{bmatrix} h_{11} & h_{12} \\ h_{21} & h_{22} \end{bmatrix} \tag{9.8}$$

Clearly h_{11} behaves as an inpedance, h_{22} as an admittance; h_{12} and h_{21} are dimensionless. The equivalent circuit corresponding to (9.7) is shown in Fig. 9.4.

Obviously, the parameters can be experimentally determined since

$$h_{11} = \left(\frac{V_1}{I_1}\right)_{V_2=0}, \quad h_{21} = \left(\frac{I_2}{I_1}\right)_{V_2=0},$$

$$h_{12} = \left(\frac{V_1}{V_2}\right)_{I_1=0}, \quad h_{22} = \left(\frac{I_2}{V_2}\right)_{I_1=0} \tag{9.9}$$

9.1.4 Transmission or *a* parameters

In the case where we wish to consider transmission of an input signal from one network to another, it is useful, for reasons which will become obvious, to consider V_2 and I_2 as independent but with I_2 *flowing in the same direction as* I_1 (i.e. *opposite* to the direction shown in Fig. 9.1). Then

$$V_1 = a_{11}V_2 + a_{12}I_2$$
$$I_1 = a_{21}V_2 + a_{22}I_2 \qquad (9.10)$$

so that

$$\begin{bmatrix} V_1 \\ I_1 \end{bmatrix} = [a] \begin{bmatrix} V_2 \\ I_2 \end{bmatrix} \quad \text{where} \quad [a] = \begin{bmatrix} a_{11} & a_{12} \\ a_{21} & a_{22} \end{bmatrix} \qquad (9.11)$$

where a_{11} and a_{22} are dimensionless, a_{12} is an 'impedance' and a_{21} is an 'admittance'. Putting $I_2 = 0$ or $V_2 = 0$ gives

$$a_{11} = \left(\frac{V_1}{V_2}\right)_{I_2=0}, \quad a_{21} = \left(\frac{I_1}{V_2}\right)_{I_2=0},$$

$$a_{12} = \left(\frac{V_1}{I_2}\right)_{V_2=0}, \quad a_{22} = \left(\frac{I_1}{I_2}\right)_{V_2=0} \qquad (9.12)$$

It is *not* possible to represent (9.10) by an equivalent circuit.

9.2 PARAMETER CONVERSION

It should be clear that there will be a unique relationship between any two sets of above parameters. Thus manipulation of (9.7) gives

$$V_1 = \frac{h_{11}h_{22} - h_{12}h_{21}}{h_{22}} I_1 + \frac{h_{12}}{h_{22}} I_2 = \frac{\Delta_h}{h_{22}} I_1 + \frac{h_{12}}{h_{22}} I_2 \qquad (9.13)$$

$$V_2 = -\frac{h_{21}}{h_{22}} I_1 + \frac{1}{h_{22}} I_2$$

where $\Delta_h = h_{11}h_{22} - h_{12}h_{21} = \|h\|$. Comparison of (9.13) and (9.1) shows that

$$z_{11} = \Delta_h/h_{22}, \; z_{12} = h_{12}/h_{22}, \; z_{21} = -h_{21}/h_{22} \text{ and } z_{22} = h_{22}^{-1}$$

Similar manipulation in the other cases gives Table 9.1 in which the rows show the conversion from the left-hand set of elements into each other set. (In verifying this table, the reader should remember that, in the case of the a parameters, the output current, I_2, is taken in the opposite sense to that for the z, y and h parameters.)

Table 9.1 Parameter conversion table

z_{11}	z_{12}	y_{22}/Δ_y	$-y_{12}/\Delta_y$	Δ_h/h_{22}	h_{12}/h_{22}	a_{11}/a_{21}	Δ_a/a_{21}
z_{21}	z_{22}	$-y_{21}/\Delta_y$	y_{11}/Δ_y	$-h_{21}/h_{22}$	$1/h_{22}$	$1/a_{21}$	a_{22}/a_{21}
y_{11}	y_{12}	$1/h_{11}$	$-h_{12}/h_{11}$	a_{22}/a_{12}	$-\Delta_a/a_{12}$	z_{22}/Δ_z	$-z_{12}/\Delta_z$
y_{21}	y_{22}	h_{21}/h_{11}	Δ_h/h_{11}	$-1/a_{12}$	a_{11}/a_{12}	$-z_{21}/\Delta_z$	z_{11}/Δ_z
h_{11}	h_{12}	a_{12}/a_{22}	Δ_a/a_{22}	Δ_z/z_{22}	z_{12}/z_{22}	$1/y_{11}$	$-y_{12}/y_{11}$
h_{21}	h_{22}	$-1/a_{22}$	a_{21}/a_{22}	$-z_{21}/z_{22}$	$1/z_{22}$	y_{21}/y_{11}	Δ_y/y_{11}
a_{11}	a_{12}	z_{11}/z_{21}	Δ_z/z_{21}	$-y_{22}/y_{21}$	$-1/y_{21}$	$-\Delta_h/h_{21}$	$-h_{11}/h_{21}$
a_{21}	a_{22}	$1/z_{21}$	z_{22}/z_{21}	$-\Delta_y/y_{21}$	$-y_{11}/y_{21}$	$-h_{22}/h_{21}$	$-1/h_{21}$

Note: $\Delta_a = a_{11}a_{22} - a_{12}a_{21}$, etc.

9.3 THE LOADED TWO-PORT NETWORK

A common configuration of a two-port network is that it should have a source of voltage (or current) connected to its input and a load connected across its output as shown in Fig. 9.5.

It can be seen that

$$V_1 = V_s - I_1 Z_s \tag{9.14}$$

and

$$V_2 = -I_2 Z_L \tag{9.15}$$

If the network is represented by its z parameters, substitution from

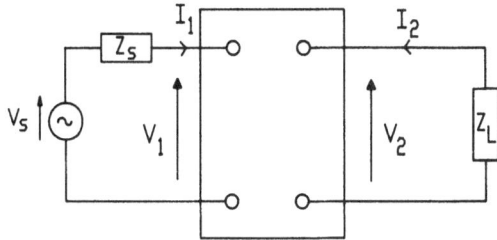

Fig. 9.5 A loaded two-port network with a voltage source (in the Thévenin representation) across it.

(9.14) and (9.15) in (9.1) gives

$$\begin{bmatrix} V_s \\ 0 \end{bmatrix} = \begin{bmatrix} z_{11} + Z_s & z_{12} \\ z_{21} & z_{22} + Z_L \end{bmatrix} \begin{bmatrix} I_1 \\ I_2 \end{bmatrix} \qquad (9.16)$$

whence

$$I_1 = \frac{V_s(z_{22} + Z_L)}{\Delta} \quad \text{and} \quad I_2 = -\frac{V_s z_{21}}{\Delta} \qquad (9.17)$$

where

$$\Delta = \begin{vmatrix} z_{11} + Z_s & z_{12} \\ z_{21} & z_{22} + Z_L \end{vmatrix} = (z_{11} + Z_s)(z_{22} + Z_L) - z_{12} z_{21}$$

If I_s is the current from the source treated as a Norton's equivalent circuit, the $V_s = I_s Z_s$ and the current gain is given from the second of (9.17) as

$$A_i = I_2/I_s = -Z_s z_{21}/\Delta \qquad (9.18)$$

which, for $Z_s \to \infty$, reduces to

$$A_i = \frac{-z_{21}}{z_{22} + Z_L} = \frac{I_2}{I_1} \qquad (9.19)$$

The ideal current coupling conditions, $Z_s \to \infty$ and $Z_L \to 0$, give a current gain

$$(A_i)_{\text{ideal}} = \frac{-z_{21}}{z_{22}} \qquad (9.20)$$

The voltage amplification, $A_v = V_2/V_s$, is obtained by substituting for I_2 from (9.17) in (9.15) when

$$A_v = \frac{z_{21} Z_L}{\Delta} \qquad (9.21)$$

The ideal gain is obtained for a low impedance source ($Z_s \to 0$) with an infinite load Z_L when

$$(A_v)_{\text{ideal}} = \frac{z_{21}}{z_{11}} \qquad (9.22)$$

which is just $(V_2/V_1)_{I_2=0}$.

From the two equations (9.17)

$$I_2 = -\frac{z_{21}}{z_{22} + Z_L} I_1$$

and, substituting in (9.1), this gives

$$V_1 = z_{11}I_1 - z_{12}\frac{z_{21}}{z_{22} + Z_L}I_1$$

Thus the input impedance given by V_1/I_1 is

$$Z_{IN} = z_{11} - \frac{z_{12}z_{21}}{z_{22} + Z_L} \tag{9.23}$$

From (9.16)

$$V_s = (z_{11} + Z_s)I_1 + z_{12}I_2$$

and

$$I_1 = (V_s - z_{12}I_2)/(z_{11} + Z_s)$$

Substituting for I_1 to get V_2 from (9.1) gives

$$V_2 = \frac{z_{21}(V_s - z_{12}I_2)}{z_{11} + Z_s} + I_2 z_{22}$$

$$= \frac{z_{21}V_s}{z_{11} + Z_s} - (-I_2)\left(z_{22} - \frac{z_{12}z_{21}}{z_{11} + Z_s}\right) \tag{9.24}$$

This corresponds to the Thévenin generator of Fig. 9.6 if

$$V_o = \frac{z_{21}V_s}{z_{11} + Z_s} = (A_v)_{oc}V_s$$

where $(A_v)_{oc}$ is the open-circuit voltage amplification from (9.21) and the output impedance is

$$Z_{OUT} = z_{22} - \frac{z_{12}z_{21}}{z_{11} + Z_s} \tag{9.25}$$

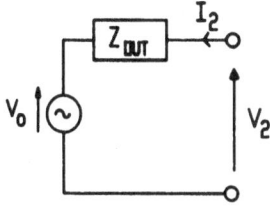

Fig. 9.6 Thévenin equivalent generator for (9.15).

Table 9.2 Amplification of current and voltage, and input and output impedances, in different parametric representations

	z parameters	y parameters	h parameters	a parameters
$(A_i)_{sc}$	$\dfrac{-Z_s z_{21}}{\Delta_z + z_{22}Z_s}$	$\dfrac{y_{21}}{-y_{11} + Y_s}$	$\dfrac{h_{21}Z_s}{h_{11} + Z_s}$	$\dfrac{Z_s}{a_{22}Z_s + a_{12}}$
$(A_i)_{ideal}$	$\dfrac{-z_{21}}{z_{22}}$	$\dfrac{y_{21}}{y_{11}}$	h_{21}	$\dfrac{1}{a_{22}}$
$(A_v)_{oc}$	$\dfrac{z_{21}}{z_{11} + Z_s}$	$\dfrac{-y_{21}Y_s}{\Delta_y + y_{22}Y_s}$	$\dfrac{-h_{21}}{\Delta_h + h_{22}Z_s}$	$\dfrac{1}{a_{21}Z_s + a_{11}}$
$(A_v)_{ideal}$	$\dfrac{z_{21}}{z_{11}}$	$\dfrac{-y_{21}}{y_{22}}$	$\dfrac{-h_{21}}{\Delta_h}$	$\dfrac{1}{a_{11}}$
Z_{IN}	$\dfrac{\Delta_z + z_{11}Z_L}{z_{22} + Z_L}$	$\dfrac{y_{22} + Y_L}{\Delta_y + y_{11}Y_L}$	$\dfrac{h_{11} + \Delta_h Z_L}{1 + h_{22}Z_L}$	$\dfrac{a_{11}Z_L + a_{12}}{a_{21}Z_L + a_{22}}$
Z_{OUT}	$\dfrac{\Delta_z + z_{22}Z_s}{z_{11} + Z_s}$	$\dfrac{y_{11} + Y_s}{\Delta_y + y_{22}Y_s}$	$\dfrac{h_{11} + Z_s}{\Delta_h + h_{22}Z_s}$	$\dfrac{a_{12} + a_{22}Z_s}{a_{21}Z_s + a_{11}}$

Note: $Y_s = Z_s^{-1}$, $\Delta_z = z_{11}z_{22} - z_{12}z_{21}$, etc.

(It should be noted that, by using (5.16), i.e. $Z_j^{IN} = \Delta/\Delta_{jj}$ with $j = 1$, (9.23) could have been derived directly from (9.16) since, to find Z_{IN}, Z_s is not included with z_{11}. Similarly, (9.25) follows from (5.18) and (9.16) when Z_L is not included with z_{22} and $Z_k^{OUT} = \Delta/\Delta_{kk}$ with $k = 2$.)

The loaded two-port network may also be described in terms of y, h and a parameters and Table 9.2 lists the gain, input and output for the four cases.

9.4 TWO-PORT NETWORKS CONNECTED IN CASCADE

Figure 9.7 shows two-port networks connected in cascade where, for obvious reasons, the current I_2 which flows from the first network to the second is shown in the same direction as I_1. This situation can be most easily expressed in terms of the sets of a parameters,

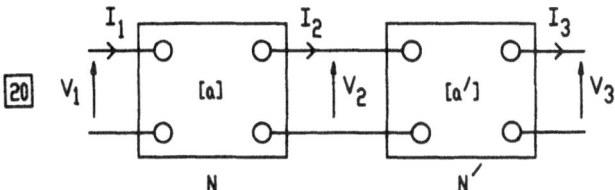

Fig. 9.7 Two-port networks N and N' in cascade.

[a] and [a'] for the networks N and N'. From (9.11)

$$\begin{bmatrix} V_1 \\ I_1 \end{bmatrix} = [a] \begin{bmatrix} V_2 \\ I_2 \end{bmatrix}$$

and

$$\begin{bmatrix} V_2 \\ I_2 \end{bmatrix} = [a'] \begin{bmatrix} V_3 \\ I_3 \end{bmatrix}$$

so that

$$\begin{bmatrix} V_1 \\ I_1 \end{bmatrix} = [a][a'] \begin{bmatrix} V_3 \\ I_3 \end{bmatrix} = [a_T] \begin{bmatrix} V_3 \\ I_3 \end{bmatrix} \qquad (9.26)$$

where $[a_T]$ is the total transmission matrix.

This process, in the a parameter representation, can be continued indefinitely and is illustrated below by considering the transmission matrices for some simple systems as illustrated in Fig. 9.8.

1. In Fig. 9.8(a) the voltages and currents are related by

$$V_1 = V_2 + ZI_2$$

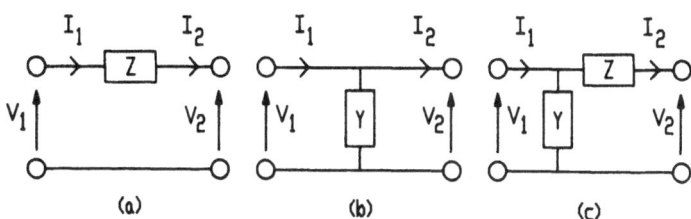

Fig. 9.8 Three simple two-port networks.

and
$$I_1 = I_2$$

so that, by analogy with (9.10), $a_{11} = 1$, $a_{12} = Z$, $a_{21} = 0$ and $a_{22} = 1$, or

$$[a] = \begin{bmatrix} 1 & Z \\ 0 & 1 \end{bmatrix} \qquad (9.27)$$

2. For Fig. 9.8(b)

$$V_1 = V_2$$

and

$$I_1 = YV_2 + I_2$$

so that $a_{11} = 1$, $a_{12} = 0$, $a_{21} = Y$ and $a_{22} = 1$, or

$$[a] = \begin{bmatrix} 1 & 0 \\ Y & 1 \end{bmatrix} \qquad (9.28)$$

3. In Fig. 9.8(c)

$$V_1 = V_2 + ZI_2$$
$$I_1 = YV_1 + I_2 = Y(V_2 + ZI_2) + I_2 = YV_2 + (1 + YZ)I_2$$

so that

$$[a] = \begin{bmatrix} 1 & Z \\ Y & 1 + YZ \end{bmatrix} \qquad (9.29)$$

On inspection, it is seen that the network in Fig. 9.8(c) is simply produced by cascading the previous two networks (with Fig. 9.8(b) first) which is confirmed by applying (9.26) to (9.28) and (9.27) to give

$$\begin{bmatrix} 1 & 0 \\ Y & 1 \end{bmatrix} \begin{bmatrix} 1 & Z \\ 0 & 1 \end{bmatrix} = \begin{bmatrix} 1 & Z \\ Y & 1 + YZ \end{bmatrix}$$

(It is easily shown that no such consistent relation can be obtained by using the convention for the sign of I_2 which applies for z, y and h parameters.)

9.5 CHARACTERISTIC IMPEDANCE

For any two-port network there is a *characteristic impedance*, Z_o, such that, if it is used to terminate the network, the input impedance

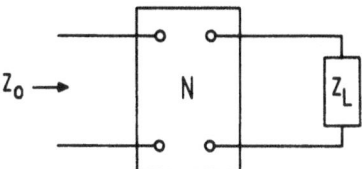

Fig. 9.9 Two-port network N terminated by its characteristic impedance.

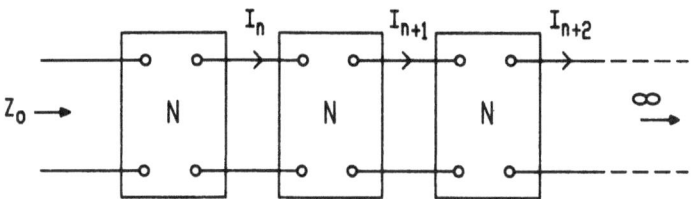

Fig. 9.10 Infinite chain or ladder of identical two-port networks of type N in cascade.

of the network is also Z_0. This situation is illustrated in Fig. 9.9. It should be obvious that, if an infinite number of identical networks of type N are connected in cascade, as indicated in Fig. 9.10, the input impedance of the system will be Z_0. This is true since the input impedance cannot be changed by adding another network of type N at the beginning and that can only be so if the terminating impedance which the new network sees is Z_0. Because of this property, Z_0 is also known as the *iterative* impedance. (If the network N is symmetrical, i.e. its two ports may be interchanged, then the characteristic impedance will be the same for both ports. However, if the network is asymmetrical, there will be two different characteristic impedances for the two ports.)

9.6 PROPAGATION CONSTANT

In Fig. 9.10, the successive currents for the nth, $(n + 1)$th and $(n + 2)$th sections of the ladder are shown as I_n, I_{n+1}, I_{n+2}. Clearly, since the ladder is infinite

$$\frac{I_n}{I_{n+1}} = \frac{I_{n+1}}{I_{n+2}} = \cdots = e^{\gamma} \tag{9.30}$$

where γ is called the *propagation constant* of the network N.

The quantity e^γ is simply the ratio between the input and output currents of the network when it is terminated in its characteristic impedance; since I_n and I_{n+1} are not necessarily in phase, γ will in general be complex. If its value is taken as

$$\gamma = \alpha + j\beta$$

where α is the *attenuation coefficient* and β is the *phase-change coefficient* (since it determines the phase shift introduced by the network), then the magnitude of the current is changed by the network by the ratio

$$\frac{|I_n|}{|I_{n+1}|} = e^\alpha \quad \text{or} \quad \alpha = \log_e\left(\frac{|I_n|}{|I_{n+1}|}\right) \tag{9.31}$$

where the unit of α is the *neper*.

If the network is being used as an attenuator terminated by its characteristic impedance, then the ratio of the power-in to the power-out will be

$$P_{\text{IN}}/P_{\text{OUT}} = |I_n|^2/|N_{n+1}|^2 = e^{2\alpha}$$

and the loss in the attenuator will be, in decibels,

$$\begin{aligned} 10\log_{10}(P_{\text{IN}}/P_{\text{OUT}}) &= 20\alpha\log_{10} e \text{ dB} \qquad \text{per section} \\ &= 8.686\alpha \text{ dB} \end{aligned} \tag{9.32}$$

(Thus 1 neper $\equiv$ 8.686 dB.)

It should be obvious that the insertion loss represented by (9.32) is an important concept in network theory.

9.7 SYMMETRICAL T AND Π NETWORKS

Any ladder of identical two-port networks can be represented by a cascade of identical symmetrical T or Π networks as is seen by inspection in Fig. 9.11. (The reader can easily demonstrate this principle by applying $\lambda - \triangle$ transformations to any sequence of identical networks they like to construct.)

9.7.1 Symmetric T network

Consider a symmetric T section from Fig. 9.11(a) and let it have a load Z_L as shown in Fig. 9.12. The input impedance is given by

$$Z_{\text{IN}} = \frac{z_1}{2} + \frac{z_2(z_1/2 + Z_L)}{z_1/2 + z_2 + Z_L} \tag{9.33}$$

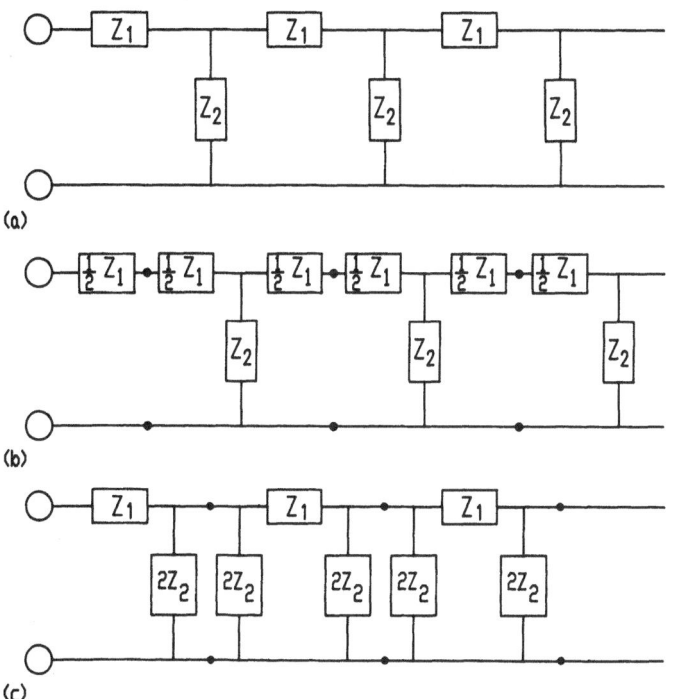

Fig. 9.11 Different but equivalent representations of a ladder of identical networks: (a) basic network, (b) symmetric T representation, and (c) symmetric Π representation.

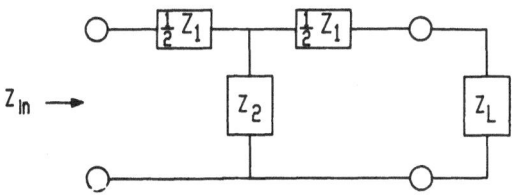

Fig. 9.12 Symmetric T with load Z_L.

If Z_L is the characteristic impedance Z_{0T}, so is Z_{IN} (see definition in section 9.5) and (9.33) then gives

$$Z_{0T} = (z_1 z_2 + z_1^2/4)^{1/2} = (z_1 z_2)^{1/2} \left(1 + \frac{z_1}{4 z_2}\right)^{1/2} \qquad (9.34)$$

The open-circuit input impedance is given by putting $Z_L = \infty$ in

(9.33),

$$Z_{oc} = z_1/2 + z_2 \tag{9.35}$$

while the short-circuit impedance $(Z_L = 0)$ is

$$Z_{sc} = \frac{z_1 z_2 + z_1^2/4}{z_1/2 + z_2} \tag{9.36}$$

Hence

$$Z_{oc} Z_{sc} = (z_1 z_2 + z_1^2/4) = Z_{0T}^2$$

or

$$Z_{0T} = (Z_{oc} Z_{sc})^{1/2} \tag{9.37}$$

9.7.2 Symmetric Π network

If the impedances in a Π network section of Fig. 9.11 are represented by admittances so that $Y_1 = Z_1^{-1}$, $Y_2 = Z_2^{-1}$ and $Y_2/2 = (2Z_2)^{-1}$ as shown in Fig. 9.13, where the load Z_L is replaced by Y_L^{-1}, the input admittance Y_{IN} can be written

$$Y_{IN} = \frac{Y_2}{2} + \frac{Y_1(Y_2/2 + Y_L)}{Y_1 + Y_2/2 + Y_L} \tag{9.38}$$

The characteristic admittance, obtained by putting $Y_{IN} = Y_L = Y_{0\pi}$ in (9.38), is

$$Y_{0\pi} = (Y_1 Y_2 + Y_2^2/4)^{1/2} \tag{9.39}$$

The open- and short-circuit admittances obtained by putting $Y_L = 0$ and $Y_L = \infty$ respectively in (9.38) are

$$Y_{oc} = \frac{Y_1 Y_2 + Y_2^2/4}{Y_1 + Y_2/2} \tag{9.40}$$

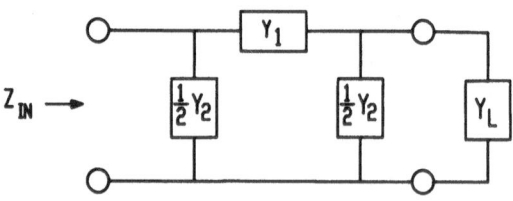

Fig. 9.13 Symmetric Π with load admittance Y_L.

and

$$Y_{sc} = Y_1 + Y_2/2 \tag{9.41}$$

whence

$$Y_{oc} Y_{sc} = Y_1 Y_2 + Y_2^2/4 = Y_{0\pi}^2 \tag{9.42}$$

In terms of the impedances, (9.39) and (9.42) give

$$Z_{0\pi} = \frac{(z_1 z_2)^{1/2}}{(1 + z_1/4z_2)^{1/2}} \tag{9.43}$$

or

$$Z_{0\pi} = (Z_{oc} Z_{sc})^{1/2} \tag{9.44}$$

(as (9.37) for the T representation) and, combining (9.34) and (9.43),

$$Z_{0\pi} Z_{0T} = z_1 z_2 \tag{9.45}$$

9.7.3 The half-section or L section

For the purpose of matching a symmetric Π section to the equivalent T section or vice versa it is possible to use the half- (or L) section obtained by dividing the Π of Fig. 9.13, or the T of Fig. 9.12, to give Fig. 9.14(a).

For any network, it is possible to find two impedances Z_A and Z_B which are 'image' impedances in that, if one terminal is loaded with Z_B, the other has an input impedance Z_A, and vice versa, as is shown for the half-section in Figs 9.14(b) and 9.14(c). (Of course, if the network were symmetric, there would be the relation $Z_A = Z_B = Z_0$.)

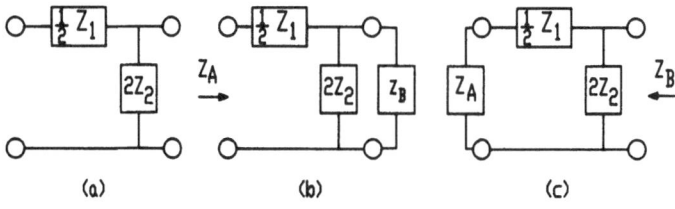

Fig. 9.14 (a) Half-section for T or Π, (b) half-section with right-hand load Z_B giving input impedance Z_A, and (c) half-section with left-hand load Z_A giving input impedance from the right as Z_B. (Z_A and Z_B are known as 'image impedances'.)

From Fig. 9.14(b)

$$Z_A = \frac{z_1}{2} + \frac{2z_1 Z_B}{2z_2 + Z_B} \qquad (9.46)$$

while, from Fig. 9.14(c),

$$Z_B = \frac{2z_2(Z_A + z_1/2)}{2z_2 + Z_A + z_1/2} \qquad (9.47)$$

Substituting from (9.47) into (9.46) and re-arranging gives

$$Z_A = (z_1 z_2 + z_1^2/4)^{1/2} = Z_{0T} \qquad (9.48)$$

and, similarly,

$$Z_B = \left(\frac{z_1 z_2}{1 + z_1/4z_2}\right)^{1/2} = Z_{0\pi} \qquad (9.49)$$

Because the image impedances at the two ends of the half-section are equal to the characteristic impedances of the corresponding T and Π sections respectively, such a half-section can be used to match a T to a Π and vice versa. (The iterative impedances of the half-section, which is asymmetrical, are different depending on which end is used as the input.)

9.8 LADDER NETWORKS

The ladder network of Fig. 9.11(a) is shown in Fig. 9.15 with the currents in the $(n-1)$th, nth and $(n+1)$th meshes set equal to I_{n-1}, I_n and I_{n+1}. The mesh equation corresponding to the nth mesh is

$$I_n(Z_1 + 2z_2) - I_{n-1}z_2 - I_{n+1}z_2 = 0$$

from which, on dividing by $2Z_2 I_n$,

$$\frac{I_{n-1}}{2I_n} + \frac{I_{n+1}}{2I_n} = \frac{z_1 + 2z_2}{2z_2} \qquad (9.50)$$

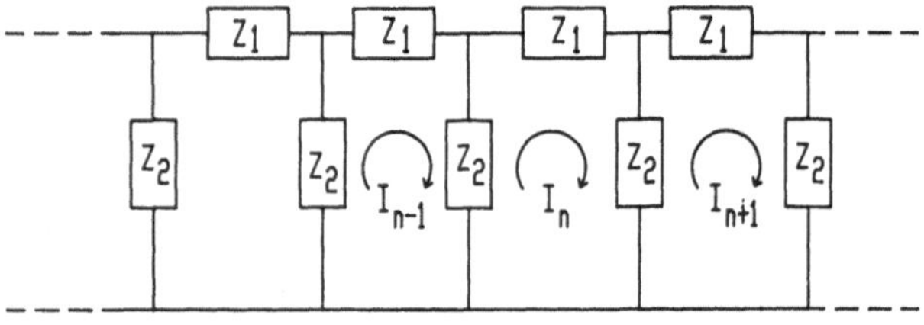

Fig. 9.15 Ladder network with mesh currents shown.

Since each section in the chain is equivalent, it is possible to write

$$\frac{I_{n-1}}{I_n} = \frac{I_n}{I_{n+1}} = e^{\gamma}$$

where γ is the propagation constant (see section 9.6) and so (9.50) becomes

$$\tfrac{1}{2}(e^{\gamma} + e^{-\gamma}) = \cosh \gamma = 1 + \frac{z_1}{2z_2} \qquad (9.51)$$

which expression must be valid for both the symmetrical T and Π networks into which the ladder subdivides.

For the symmetrical T network, substitution from (9.35) into (9.51) gives

$$\cosh \gamma = Z_{oc}/z_2 \qquad (9.52)$$

Also, putting $\sinh^2\gamma = \cosh^2\gamma - 1 = (z_1 z_2 + z_1^2/4)z_2^2$ and substituting from (9.34), gives

$$\sinh \gamma = Z_{0T}/z_2 \qquad (9.53)$$

Then, from (9.52) and (9.53), and using (9.37),

$$\tanh \gamma = Z_{0T}/Z_{oc} = (Z_{sc}Z_{oc})^{1/2}/Z_{oc} = (Z_{sc}/Z_{oc})^{1/2} \qquad (9.54)$$

If experiment (or calculation) enables $\tanh \gamma$ to be expressed as

$$\tanh \gamma = A + jB$$

and $\gamma = \alpha + j\beta$ (section 9.6), then expansion of $\tanh(\alpha + j\beta)$ gives

$$\begin{aligned} \tanh 2\alpha &= 2A/(1 + A^2 + B^2) \\ \tan 2\beta &= 2B/(1 - A^2 - B^2) \end{aligned} \qquad (9.55)$$

so that α and β can be found from standard tables.

(Because of the nature of the infinite ladder network, (9.54) must be equally valid for the equivalent Π network: it is left as an exercise for the reader to prove this.)

9.9 FILTERS

One of the principal uses of ladder networks of identical sections is as filters. A *filter* is a device which, ideally, suppresses certain frequencies (the *stop-band(s)*) while leaving other frequencies (the *pass-band(s)*) virtually unattenuated. The simplest filter structures, referred to as prototype filters or *constant-k filters*, consist of ladders of the form of Fig. 9.11 in which the product

$$Z_1 Z_2 = k^2 \qquad (9.56)$$

is independent of frequency. (It will be seen that $k = R_0$, the *design resistance* of the filter.) Other, more sophisticated filters can be developed from the constant-k design.

The commonest types of filter are the *low-pass, high-pass, band-pass* and *band-stop* filters: discussion will be limited here to constant-k low-pass and high-pass filters.

9.9.1 Low-pass constant-k filters

The ideal low-pass filter will have a pass-band of zero attenuation from dc to f_c, a cut-off frequency, and a stop-band with infinite attenuation above f_c. Figure 9.16 shows the corresponding T and Π sections for a simple low-pass constant-k filter which has $Z_1 = j\omega L$ and $Z_2 = (j\omega C)^{-1}$.

The product $Z_1 Z_2 = L/C$ which corresponds to a constant-k filter with the design impedance

$$R_0 = (L/C)^{1/2} \qquad (9.57)$$

From (9.51)

$$\cosh \gamma = 1 + \frac{Z_1}{2Z_2} = 1 - \tfrac{1}{2}\omega^2 LC \qquad (9.58)$$

But $\cosh \gamma = \cosh(\alpha + j\beta) = \cosh \alpha \cos \beta + j \sinh \alpha \sin \beta$ so that

$$\sinh \alpha \sin \beta = 0 \qquad (9.59)$$

and

$$\cosh \alpha \cos \beta = 1 - \tfrac{1}{2}\omega^2 LC \qquad (9.60)$$

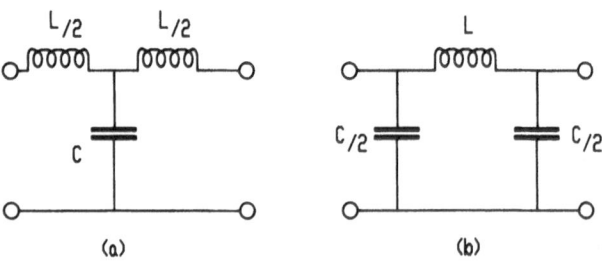

Fig. 9.16 (a) T section and (b) Π section for a simple low-pass, constant-k filter.

If, in the pass-band, $\alpha = 0$ (i.e. zero attenuation), (9.59) is satisfied and $\cosh \alpha = 1$ so that

$$\cos \beta = 1 - \tfrac{1}{2}\omega^2 LC \qquad (9.61)$$

But $-1 \leqslant \cos \beta \leqslant 1$ so that (9.61) can only be true for $0 \leqslant \omega \leqslant 2/\sqrt{LC}$ or $0 \leqslant f \leqslant 1/\pi\sqrt{LC}$. (Outside this range, the condition $\alpha = 0$ cannot be satisfied.) There will thus be zero attenuation in a pass-band up to the cut-off frequency $f_c = 1/\pi\sqrt{LC}$.

The phase shift at a frequency f in the pass-band is given by (9.61) as

$$\beta = \cos^{-1}[1 - 2(f/f_c)^2] \qquad (9.62)$$

which varies from 0 to π as the frequency varies from zero to f_c. For $f > f_c$, $\alpha \neq 0$ and (9.59) requires $\sin \beta = 0$.

The characteristic impedance of the T (or Π) section will vary with frequency. For the T section, (9.34) gives

$$Z_{0T} = (Z_1 Z_2)^{1/2}(1 + Z_1/4Z_2)^{1/2}$$

which becomes

$$Z_{0T} = \left(\frac{L}{C}\right)^{1/2}[1 - (f/f_c)^2]^{1/2} \qquad (9.63)$$

and the characteristic impedance varies from $(L/C)^{1/2}$ for $f = 0$ to zero for $f = f_c$. Similarly, from (9.45), the Π section characteristic impedance varies as

$$Z_{0\pi} = Z_1 Z_2/Z_{0T} = (L/C)^{1/2}[1 - (f/f_c)^2]^{-1/2} \qquad (9.64)$$

becoming $(L/C)^{1/2}$ at $f = 0$ and infinity at $f = f_c$.

Since (9.62) shows $\beta = \pi$ at the cut-off frequency, and $\sin \beta = 0$ is required in the stop-band (since $\alpha \neq 0$), it is obvious to take $\beta = \pi$ and $\cos \beta = -1$ throughout the stop-band. Then (9.60) gives

$$\alpha = \cosh^{-1}[2(f/f_c)^2 - 1] \qquad (9.65)$$

The variation of α, β, Z_{0T} and $Z_{0\pi}$ with frequency is shown in Figs 9.17(a) and 9.17(b).

9.9.2 The high-pass constant-k filter

The simplest T and Π representations of the sections for a constant-k high-pass filter are shown in Fig. 9.18: since the filter is required to

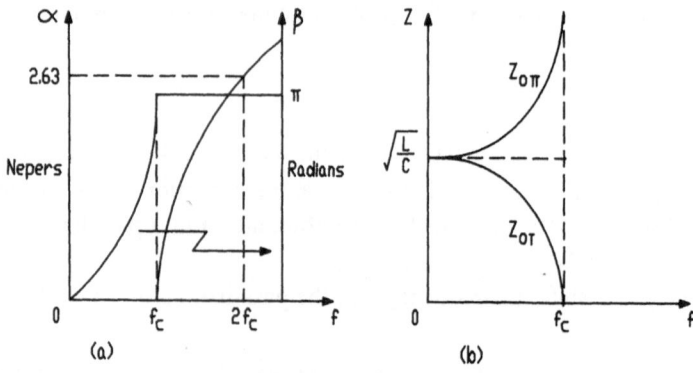

Fig. 9.17 (a) Variation of α and β with frequency, and (b) variation of $Z_{0\Pi}$ and Z_{0T} with frequency, for the low-pass case.

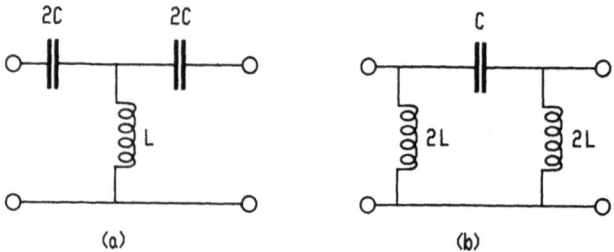

Fig. 9.18 (a) T section and (b) Π section for a high-pass filter.

give attenuation below a frequency f_c and none above (the opposite of the low-pass filter), it is obvious to invert the components of the low-pass device.

Once again

$$k = R_0 = (L/C)^{1/2}$$

is independent of frequency. From (9.51)

$$\cosh \gamma = 1 - (2\omega^2 LC)^{-1} \qquad (9.66)$$

which, by analogy with (9.59) and (9.60), gives

$$\cosh \alpha \cos \beta = 1 - (2\omega^2 LC)^{-1}$$
$$\sin \alpha \sin \beta = 0 \qquad (9.67)$$

From (9.67), $\alpha = 0$ for $\omega > \omega_c = \frac{1}{2}\sqrt{LC}$ or $f > f_c = 1/4\pi\sqrt{LC}$, which means there is a pass-band for $f_c < f < \infty$. Within the pass-band, where $\cosh\alpha = 0$, the phase shift is given by

$$\beta = \cos^{-1}[1 - (2\omega^2 LC)^{-1}] = \cos^{-1}\left[1 - 2\left(\frac{f_c}{f}\right)^2\right] \qquad (9.68)$$

which has the value $-\pi$ at f_c (and zero at $f = \infty$).

As in the case of the low-pass filter, using (9.34) and (9.45) gives

$$Z_{0T} = \left(\frac{L}{C}\right)^{1/2}\left[1 - \left(\frac{f_c}{f}\right)^2\right]^{1/2} \qquad (9.69)$$

and

$$Z_{0\pi} = \left(\frac{L}{C}\right)^{1/2}\left[1 - \left(\frac{f_c}{f}\right)^2\right]^{-1/2} \qquad (9.70)$$

which vary, respectively, from 0 to $(L/C)^{1/2}$ and ∞ to $(L/C)^{1/2}$ across the pass band.

Taking $\cos\beta$ at its limiting value of -1 for $0 \leqslant f \leqslant f_c$ (see (9.68)), (9.66) gives

$$\alpha = \cosh^{-1}[2(f_c/f)^2 - 1] \qquad (9.71)$$

Figure 9.19 shows the variation of α, β, Z_{0T} and $Z_{0\pi}$ with frequency in this case.

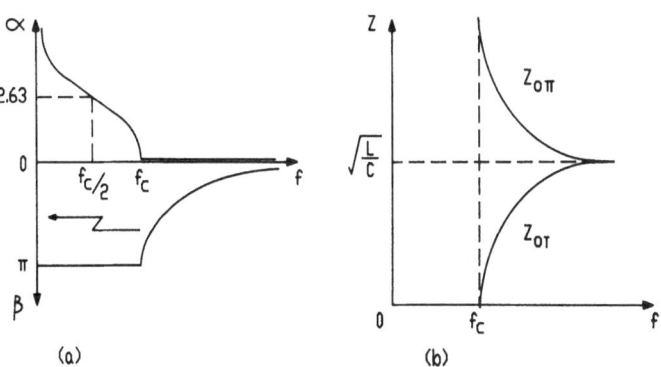

Fig. 9.19 (a) Variation of α and β with frequency, and (b) variation of $Z_{0\pi}$ and Z_{0T} with frequency, for the high-pass case.

9.9.3 Design parameters and problems

The design parameters of the prototype filters are as follows: cut-off frequency, f_c, is $1/\pi\sqrt{LC}$ (low pass), but $1/4\pi\sqrt{LC}$ (high pass); design resistance, R_0, is $(L/C)^{1/2}$ in both cases, where R_0 approximates to the characteristic impedances Z_{0T} and $Z_{0\pi}$ for frequencies well within the pass-band. The choice of f_c and R_0 fixes L and C in the network.

All prototype filters suffer two limitations.

1. The characteristic impedance varies significantly near the edge of the pass-band so that matching is difficult.
2. The attenuation coefficient rises only slowly near the cut-off frequency (although its absolute value can be increased by increasing the numbers of stages in the ladder since, with m stages, the attenuation coefficient will increase to $m\alpha$).

9.10 ATTENUATORS

An ideal attenuator is a device which will diminish the magnitude of a signal by an amount which is independent of frequency, and produce zero phase shift in the process. The simplest such attenuator will be a ladder network in which Z_1 and Z_2 are either both ideal resistors or both ideal capacitors.

Consider the resistive attenuator shown in Fig. 9.20. From (9.51)

$$\cosh\gamma = 1 + R_1/2R_2$$

Thus $\cosh\gamma$ is real, greater than unity and independent of frequency: there will be attenuation at all frequencies without phase shift. By

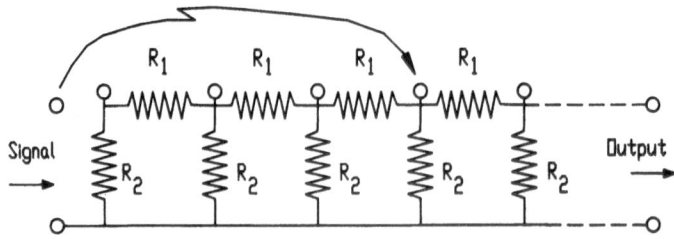

Fig. 9.20 Attenuator employing resistances. The signal may be connected at any given input point as indicated so that the attenuation can be varied.

analogy with (9.59) and (9.60),

$$\cosh \alpha \cos \beta = 1 + R_1/2R_2$$
$$\sinh \alpha \sin \beta = 0 \tag{9.72}$$

If $\sinh \alpha = 0$, which implies $\cosh \alpha = 1$, $\cos \beta = (1 + R_1/2R_2) > 1$. This is impossible, so $\sinh \alpha \neq 0$ and, consequently, $\sin \beta$ must be zero: there is no phase shift. Since $\cos \beta = 1$, (9.72) requires

$$\alpha = \cosh^{-1}(1 + R_1/2R_2) \tag{9.73}$$

which is independent of frequency. (If Z_1 and Z_2 were due to capacitance C_1 and C_2, then again $\beta = 0$: the attenuation would be $\alpha = \cosh^{-1}(1 + C_2/2C_1)$.)

The ladder network attenuator is superior to the simple potential divider since, with a proper terminating impedance, the load presented to the signal is independent of the number of sections used and, therefore, is independent of the actual attenuation: this is not the case with a potential divider unless it has a very high terminating impedance (i.e. the detector) into which it feeds.

9.11 TRANSMISSION LINES

Transmission lines, such as telephone wires or coaxial cables, consist of two parallel conductors which carry an electrical signal from a generator to a load. The behaviour of such lines can, in principle, be worked out exactly by application of Maxwell's electromagnetic equations. However, in this section, an approximate treatment will be given in which a transmission line is considered as being a uniform ladder network with *distributed* rather than *lumped* impedances in the series and shunt arms.

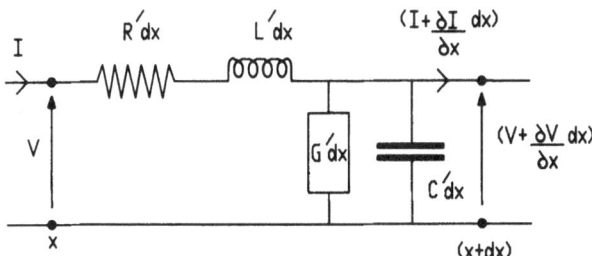

Fig. 9.21 Short length dx of a transmission line represented as a uniform ladder network with distributed R', L', G' and C'.

If the series resistance and series inductance of a line are taken to be R' and L' per unit length, and the parallel conductance and parallel capacitance to be G' and C' per unit length, then the short length dx of the line, which lies at a distance between x and $x + dx$ from a reference point $x = 0$, can be represented by the circuit shown in Fig. 9.21. The voltages at x and $x + dx$ can be taken as V and $V + (\partial V/\partial x)\,dx$, while the corresponding currents will be I and $I + (\partial I/\partial x)\,dx$.

Application of Kirchhoff's voltage law to the network between x and $x + dx$ gives

$$V - \left(V + \frac{\partial V}{\partial x}\,dx \right) = (R'\,dx)I + (L'\,dx)\frac{\partial I}{\partial t}$$

or

$$-\frac{\partial V}{\partial x} = R'I + L'\frac{\partial I}{\partial t} \qquad (9.74)$$

Similarly, application of Kirchhoff's current law gives

$$I - \left(I + \frac{\partial I}{\partial x}\,dx \right) = (G'\,dx)V + (C'\,dx)\frac{\partial V}{\partial t}$$

or

$$-\frac{\partial I}{\partial x} = G'V + C'\frac{\partial V}{\partial t} \qquad (9.75)$$

(In these equations, $\partial V/\partial x \equiv (\partial V/dx)_{t=\text{constant}}$, $\partial V/\partial t \equiv (\partial V/\partial t)_{x=\text{constant}}$.) Suppose that the voltage and the current both vary with time so that, at a particular point x, $V = V_0\,e^{j\omega t}$ and $I = I_0\,e^{j\omega t}$. Equations (9.74) and (9.75) then become, on eliminating $e^{j\omega t}$,

$$\frac{\partial V_0}{\partial x} = -(R' + j\omega L')I_0 = -Z'I_0 \qquad (9.76)$$

and

$$\frac{\partial I_0}{\partial x} = -(G' + j\omega C')V_0 = -Y'V_0 \qquad (9.77)$$

where $Z' = R' + j\omega L'$ and $Y' = G' + j\omega C'$ represent respectively the series impedance and the shunt admittance per unit length of the line; Z' and Y' are independent of x.

Differentiating the above two equations with respect to x and eliminating gives

$$\frac{\partial^2 V_0}{\partial x^2} = Y'Z'V_0 \quad \text{and} \quad \frac{\partial^2 I_0}{\partial x^2} = Y'Z'I_0 \tag{9.78}$$

These differential equations have the solutions

$$V_0 = A\,e^{-\gamma'x} + B\,e^{\gamma'x} \quad \text{and} \quad I_0 = C\,e^{-\gamma'x} + D\,e^{\gamma'x} \tag{9.79}$$

where $\gamma' = \sqrt{Y'Z'}$, the propagation constant, is generally complex and of the form $\gamma' = \alpha' + j\beta'$.

The complete expressions for V and I are then

$$V = A\,e^{-\alpha'x}\,e^{j(\omega t - \beta'x)} + B\,e^{\alpha'x}\,e^{j(\omega t + \beta'x)}$$
$$I = C\,e^{-\alpha'x}\,e^{j(\omega t - \beta'x)} + D\,e^{\alpha'x}\,e^{j(\omega t + \beta'x)} \tag{9.80}$$

For both V and I, the first term represents a wave travelling in the positive x-direction with attenuation $e^{-\alpha'x}$, and the second term represents a wave travelling in the negative x-direction with attenuation $e^{\alpha'x}$ (x being increasingly negative). β', the phase constant of the wave, may be taken to be given by $2\pi/\lambda$ where λ is the wavelength, so that the wave velocity is $v = \omega/\beta'$.

9.11.1 Characteristic impedance

Substitution from (9.79) into (9.76) gives

$$-\gamma'A\,e^{-\gamma'x} + \gamma'B\,e^{\gamma'x} = -Z'(C\,e^{-\gamma'x} + D\,e^{\gamma'x})$$

Equating coefficients gives

$$C = \frac{\gamma'A}{Z'} = \sqrt{\frac{Y'}{Z'}}\,A = \frac{A}{Z_0} \quad \text{and} \quad D = -\frac{B}{Z_0} \tag{9.81}$$

where, using (9.76), (9.77) and (9.79),

$$Z_0 = \sqrt{\frac{Z'}{Y'}} = \left(\frac{R' + j\omega L'}{G' + j\omega C'}\right)^{1/2} \tag{9.82}$$

It is clear that Z_0 is the voltage to current ratio for both the forward wave and the reverse wave at every point on the transmission line; it is called the characteristic impedance of the line.

For a wave travelling in the positive x-direction, a forward wave, the current will reduce to zero at $x = \infty$; there will be no reverse or

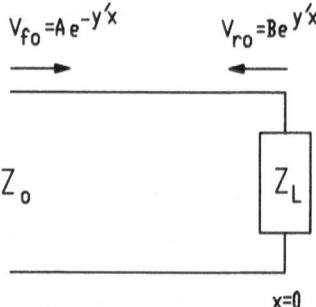

Fig. 9.22 Transmission line terminated with a load Z_L.

reflected wave generated. It follows that, at any point on an infinite line, the input impedance of the line will appear to be simply Z_0: consequently, if the line is cut off at a finite length but is terminated with a load impedance Z_0, there will likewise be no reflected wave, and the load will be said to be matched to the line.

Using (9.79) and (9.81), the relationship of I_0 and V_0 may be simplified so that

$$V_0 = A\,\mathrm{e}^{-\gamma'x} + B\,\mathrm{e}^{\gamma'x} \quad \text{and} \quad I_0 Z_0 = A\,\mathrm{e}^{-\gamma'x} - B\,\mathrm{e}^{\gamma'x} \quad (9.83)$$

If a load impedance Z_L is introduced at $x = 0$ as shown in Fig. 9.22, then, from (9.83),

$$Z_\text{L} = \frac{V_0(x=0)}{I_0(x=0)} = \frac{(A+B)Z_0}{A-B} = \left(\frac{1+\rho}{1-\rho}\right)Z_0 \quad (9.84)$$

where the reflection coefficient is defined as $\rho = B/A$. Measurement of ρ will give Z_L whilst, obviously, $Z_\text{L} = Z_0$ gives $\rho = 0$.

9.11.2 Low-loss approximation

In the case where the attenuation is small and it can be assumed that $R' \ll \omega L'$, $G' \ll \omega C'$, convenient approximate expressions can be obtained for α' and β'.

Since, from (9.79), $\gamma' = \alpha' + \mathrm{j}\beta' = \sqrt{(R' + \mathrm{j}\omega L')(G' + \mathrm{j}\omega C')}$ it is possible to write

$$\alpha'^2 + \beta'^2 = |\alpha' + \mathrm{j}\beta'|^2 = \sqrt{(R'^2 + \omega^2 L'^2)(G'^2 + \omega^2 C'^2)}$$

$$\simeq \omega^2 L'C' + \tfrac{1}{2}\left(\frac{R'^2 C'}{L'} + \frac{G'^2 L'}{C'}\right)$$

while

$$\alpha'^2 - \beta'^2 = \text{real part of } (\alpha' + j\beta')^2$$
$$= \text{real part of } (R' + j\omega L')(G' + j\omega C')$$
$$= R'G' - \omega^2 L'C'$$

Consequently

$$\alpha' \simeq \tfrac{1}{2}\left(R'\sqrt{\frac{C'}{L'}} + G'\sqrt{\frac{L'}{C'}} \right) \tag{9.85}$$

and

$$\beta' \simeq \omega\sqrt{L'C'}\left[1 + \frac{1}{8\omega^2}\left(\frac{G'}{C'} - \frac{R'}{L'}\right)^2 \right] \tag{9.86}$$

Thus, using (9.86), the wave velocity, $v = \omega/\beta'$, is approximately

$$v = (L'C')^{-1/2}\left[1 - \frac{1}{8\omega^2}\left(\frac{G'}{C'} - \frac{R'}{L'}\right)^2 \right] \tag{9.87}$$

while the characteristic impedance is approximately $Z_0 = (L'/C')^{1/2}$.

9.11.3 Dispersion and distortion

If the signal to be transmitted consists of a mixture of frequencies, the signal will be dispersed because the velocity will depend upon frequency (see (9.87)). There will also be distortion if the loss depends on frequency.

At high frequencies ($f > 1$ MHz), $R' \ll \omega L'$ and $G' \ll \omega C'$ and the velocity will approximate to $v = (L'C')^{-1/2}$ with no dispersion: however, the term $R'\sqrt{C'/L'}$ in (9.85) for α' will increase as $\omega^{1/2}$ owing to the skin effect.

At low frequencies, the dispersion may be serious and lead to distortion of the voice, say, in telephone conversations. Theoretically, the problem may be solved by making $Z' = a^2 Y'$ where a is a constant: this means

$$R' + j\omega L' = a^2(G' + j\omega C') \tag{9.88}$$

and

$$\gamma' = \alpha' + j\beta' = \sqrt{(R' + j\omega L')(G' + j\omega C')} = a(G' + j\omega C')$$

Then $\alpha' = aG'$ and $v = \omega/a\omega C' = 1/aC'$ would both be frequency independent and there would be no distortion. Unfortunately, in practice, the relationship of (9.88) cannot be realized since $R'/L' \gg G'/C'$: nevertheless, some improvement can be achieved by increasing either or both of L' and G'.

In low-frequency telegraph lines, L' may be increased by winding permalloy tape around one or both conductors while, in telephone cables, small coils, of dimensions less than the wavelengths to be transmitted, are added in series with the conductor at regular intervals. G' may be increased by reducing the insulation between the conductors; the loss may increase, but that can be overcome by using amplifiers (repeaters) at intervals along the line.

Two final points may be noted.

First, if the transmission line consists of two conductors—a pair of parallel wires or two concentric cylinders forming a coaxial cable—separated by free space—the velocity will approximate to $v = (L'C')^{-1/2} = (\varepsilon_0 \mu_0)^{-1/2} = c_0$, the velocity of electromagnetic radiation in free space.

Second, while the configuration of the line divided into series and shunt impedances as in Fig. 9.21 is similar to that of a lumped-element, low-pass filter (see section 9.9.1), then, because the series and shunt arms have parameters $L'\,dx$ and $C'\,dx$ respectively, the cut-off frequency will be $f_c = 1/\pi(L'C')^{1/2}\,dx$; since dx may be made as small as desired, $f_c \to \infty$ and there is no cut-off to the signal frequencies.

9.12 ARTIFICIAL DELAY LINES

Artificial delay lines or delay networks are used in a variety of circuits, i.e. for the production of pulses of short duration and in high-power modulators used in radar transmitters. However, in this section only some of the more basic delay networks will be discussed.

In section 9.11, it was shown that a signal travels along a lossless transmission line with a velocity of $(L'C')^{-1/2}$, where L' and C' are the inductance and capacitance respectively per unit length of the line. A length x of such a transmission line will therefore introduce a time delay into the signal given by

$$\tau = x/v = x(L'C')^{1/2}$$

The use of a length of cable as a delay network is, however, somewhat limited, since the length becomes inconveniently long if the required delay exceeds a fraction of a microsecond.

Distributed delay lines have been designed which reduce the length of cable required. For example, in the solenoid delay line, an inner conductor in the form of a solenoid of insulated wire wound on a PVC tube and then covered by tape is used: onto this the outer conductor is then laid. Obviously, in this device, the reduction in the length of the cable required has been achieved by increasing the capacitance and inductance associated with unit length of the cable.

In section 9.11.3 it was stated that the properties of a distributed transmission line were analogous to those of a low-pass filter network. This similarity can be extended to the use of a lumped low-pass filter network as a simple delay line. For a T section of a constant-k, low-pass filter, it has been shown in (9.63) that the characteristic impedance, Z_{0T}, is

$$Z_{0T} = \left(\frac{L}{C}\right)^{1/2} [1 - (f/f_c)^2]^{1/2} = \left(\frac{L}{C}\right)^{1/2} [1 - (\omega/\omega_c)^2]^{1/2}$$

and that the phase shift β in the pass-band, given by (9.62), is

$$\beta = \cos^{-1}[1 - 2(f/f_c)^2] = \cos^{-1}[1 - 2(\omega/\omega_c)^2]$$

where, in both the above equations, ω_c is the cut-off angular frequency equal to $2(LC)^{-1/2}$.

The time delay introduced by such a T section is given by

$$\tau = \frac{d\beta}{d\omega} = \frac{2}{\omega_c}\left[1 - \left(\frac{\omega}{\omega_c}\right)^2\right]^{-1/2} \tag{9.89}$$

Equation (9.89) shows that τ increases monotonically with frequency as is illustrated by the form of the graph showing the variation of β with frequency (Fig. 9.17(a)). However, at low frequencies where $\omega \ll \omega_c$, the delay time is approximately independent of frequency, being

$$\tau \simeq \frac{2}{\omega_c} = (LC)^{1/2} \tag{9.90}$$

Also at low frequencies, the characteristic impedance $Z_{0T} \simeq (L/C)^{1/2}$ is approximately independent of frequency.

Obviously the delay time can be increased by connecting several T sections in cascade whence, for n sections, the total delay time is given by $n\tau$.

It should be noted that, a more linear phase shift characteristic, and hence a more constant delay time, can be achieved by using a modified low-pass filter network, e.g. an m-derived filter (see F. R. Connor, NETWORKS, Edward Arnold, London, 1979).

APPENDIX A

Quantities and symbols used in the text

Quantity	Symbol	Unit	Dimensions	Equation
Electric current	I	A	I	SI unit
Electric charge	Q	C	IT	$\int I\,dt$
Electric potential (pd)	V, V_{AB}	V	$I^{-1}ML^2T^{-3}$	$\int E\cdot ds$
Electric field	E	$V\,m^{-1}$	$I^{-1}MLT^{-3}$	$-\operatorname{grad} V$
Electromotive force	V_0	V	$I^{-1}ML^2T^{-3}$	$\oint E\cdot ds$
Resistance	R	Ω	$I^{-2}ML^2T^{-3}$	VI^{-1}
Conductance	G	S	$I^2M^{-1}L^{-2}T^3$	R^{-1}
Power	P	W	ML^2T^{-3}	I^2R
Temperature coefficient of resistance	α	—	—	$\dfrac{R-R_0}{R_0(\theta-\theta_0)}$
Internal resistance	R_0	Ω	$I^{-2}ML^2T^{-3}$	$(V_0-V_{AB})I^{-1}$
Mutual inductance	M	H	$I^{-2}ML^2T^{-2}$	$V(dI/dt)^{-1}$
Self-inductance	L	H	$I^{-2}ML^2T^{-2}$	$V_{back}(dI/dt)^{-1}$
Coefficient of coupling	k	—	None	$M(L_1L_2)^{-1/2}$
Capacitance	C	F	$I^2M^{-1}L^{-2}T^4$	QV^{-1}
Time constant	τ	s	T	(3.14)
Instantaneous voltage	V	V	$I^{-1}ML^2T^{-3}$	(4.1)
Peak voltage	V_0	V	$I^{-1}ML^2T^{-3}$	(4.1)
Angular frequency	ω	$rad\,s^{-1}$	T^{-1}	(4.1)
Period	T	s	T	(4.2)
Frequency	f	Hz	T^{-1}	(4.3)
Phase difference	ϕ	$rad\,deg^{-1}$	None	(4.4)
Reactance	X	Ω	$I^{-2}ML^2T^{-3}$	(4.14)

(Contd)

(*Continued*)

Quantity	Symbol	Unit	Dimensions	Equation
Impedance	$\|Z\|$	Ω	$I^{-2}ML^2T^{-3}$	(4.17)
Complex impedance	Z	Ω	$I^{-2}ML^2T^{-3}$	(4.23)
Complex admittance	Y	S	$I^2M^{-1}L^{-2}T^3$	(4.23)
Susceptance	B	S	$I^2M^{-1}L^{-2}T^3$	(4.48)
Rms current	I_{rms}	A	I	(4.50)
Rms voltage	V_{rms}	V	$I^{-1}ML^2T^{-3}$	(4.53)
Instantaneous power	p	W	ML^2T^{-3}	(4.54)
Average power	P	W	ML^2T^{-3}	(4.55)
Apparent power	S	VA	ML^2T^{-3}	(4.55)
Power factor	$\cos\phi$	—	None	(4.55)
Reactive power	Q	var	ML^2T^{-3}	(4.56)
Complex power	S'	W	ML^2T^{-3}	(4.59)
Input impedance	Z_j^{IN}, Z_{IN}	Ω	$I^{-2}ML^2T^{-3}$	(5.25)
Output impedance	Z_k^{OUT}, Z_{OUT}	Ω	$I^{-2}ML^2T^{-3}$	(5.26)
Input admittance	Y_j^{IN}, Y_{IN}	S	$I^2M^{-1}L^{-2}T^3$	(5.19)
Output admittance	Y_k^{OUT}, Y_{OUT}	S	$I^2M^{-1}L^{-2}T^3$	(5.25)
Transfer admittance	Y_{kj}^{TRAN}	S	$I^2M^{-1}L^{-2}T^3$	(5.27)
Resonance frequency	f_0, f_p	Hz	T^{-1}	(7.2)
Q or quality factor	Q	—	None	(7.4)
Parallel dynamic impedance	Z_p	Ω	$I^{-2}ML^2T^{-3}$	(7.17)
Optimum coupling coefficient	k_0	—	None	(8.19)
Critical coupling coefficient	k_c	—	None	(8.27)
Current gain	A_i	—	None	(9.18)
Ideal current gain	$(A_i)_{ideal}$	—	None	(9.20)
Voltage amplification	A_v	—	None	(9.21)
Ideal voltage amplification	$(A_v)_{ideal}$	—	None	(9.22)
Characteristic or iterative impedance	Z_0	Ω	$I^{-2}ML^2T^{-3}$	Section 9.5

(*Contd*)

(*Continued*)

Quantity	Symbol	Unit	Dimensions	Equation
Propagation constant	γ	—	None	(9.30)
Attenuation coefficient	α	—	None	(9.31)
Phase-change coefficient	β	rad	None	(9.30)
Image impedance	Z_A, Z_B	Ω	$I^{-2}ML^2T^{-3}$	(9.46)
Design resistance	k, R_0	Ω	$I^{-2}ML^2T^{-3}$	(9.56)
Cut-off frequency	f_c	Hz	T^{-1}	(9.62)
Distributed resistance	R'	$\Omega\,m^{-1}$	$I^{-2}MLT^{-3}$	(9.74)
Distributed inductance	L'	$H\,m^{-1}$	$I^{-2}MLT^{-2}$	(9.74)
Distributed conductance	G'	$S\,m^{-1}$	$I^2M^{-1}L^{-3}T^3$	(9.75)
Distributed capacitance	C'	$F\,m^{-1}$	$I^2M^{-1}L^{-3}T^4$	(9.75)
Distributed impedance	Z'	$\Omega\,m^{-1}$	$I^{-2}MLT^{-3}$	(9.78)
Distributed admittance	Y'	$S\,m^{-1}$	$I^2M^{-1}L^{-3}T^3$	(9.78)
Propagation coefficient	γ'	m^{-1}	L^{-1}	(9.79)
Attenuation coefficient	α'	neper m^{-1}	L^{-1}	(9.80)
Phase constant	β'	rad m^{-1}	L^{-1}	(9.80)

APPENDIX B

Exercises

2 Direct current theory

2.1 A battery of emf 15 V and internal resistance $2\,\Omega$ is connected in parallel with another battery of emf 10 V and internal resistance $3\,\Omega$ so that like polarities are connected together. The combination is used to send a current through an external load of resistance $20\,\Omega$. Calculate the current flowing through each battery and the power supplied to the external load.

2.2 A dc voltage generator has a voltage across its terminals of 96 V when the current drawn from it is 200 mA. When an external load of $180\,\Omega$ is connected across its terminals the current supplied to the load is 500 mA. Calculate the emf and internal resistance of the voltage generator.

2.3 There are n cells, each of emf 2.1 V and internal resistance $0.1\,\Omega$, connected in series so that they can be charged from a dc source of emf 120 V and negligible internal resistance. If the charging current to be used is 5 A, determine (in terms of n) the additional resistance required to be connected in series with the cells in order to make this possible.

What is the maximum number of such cells which can be charged from this source using a charging current of 5 A?

If $n = 25$, what percentage of the energy delivered by the source is wasted as heat?

3 Capacitors, inductors and transients

3.1 consider the circuit shown and answer the following questions.

Appendix B

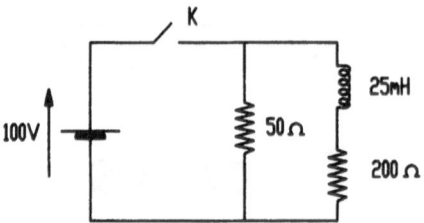

1. What is the rate of change of the current through the inductor at the instant immediately after the key K is closed?
2. What is the current through the inductor after a very long time has elapsed since the key K was closed?
3. What is the rate of change of the current through the inductor when the current through it has reached one-half of its final value?
4. What current is supplied by the source after a very long time has elapsed since the key K was closed?
5. If the key K is now opened what is the time constant for the decay of the current in the inductor?

3.2 A neon lamp, which flashes when the pd across it is 100 V and is extinguished when the pd falls to 80 V, is connected across a capacitor of capacitance 1 pF. This combination is connected in series with a resistor of resistance 1 MΩ and a battery of emf 150 V. Assuming that the duration of each neon flash is negligible, determine (a) the frequency of the flashes, and (b) the average current which flows in the circuit in the interval during which the potential increases from 80 V to 100 V.

3.3 A constant voltage of 200 V is suddenly applied to a circuit of resistance 5 Ω and self-inductance 20 H. After 5 s the voltage is suddenly increased to 300 V. What is the value of the current after a further 3 s?

3.4 The voltage across a charged capacitor falls from 100 V to 90 V in 300 s as a result of defective insulation between the plates of the capacitor. When the terminals are joined by a resistance of 10 MΩ the voltage drops from 100 V to 20 V in 120 s. Calculate the insulation resistance of the capacitance in megaohms.

3.5 In the circuit shown below, show that the instantaneous current I through the self-inductance L, t seconds after the switch S is closed, is given by

$$I = \frac{V_1}{R}(1 - e^{-R_1 t/L})$$

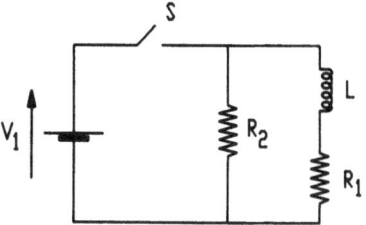

If the switch S is opened when the current flowing through L is I_0, show that the instantaneous current I' through the self-inductance t seconds after S is opened is given by

$$I' = I_0 e^{-(R_1 + R_2)t/L}$$

If $R_1 = R_2 = 1\,k\Omega$, $V_1 = 40\,V$ and L is the self-inductance of the coil of a relay which has an operating current of 30 mA and releases when the current falls to 15 mA, determine

1. the value of L if the relay is to operate 3 ms after the switch S is closed, and
2. the release time of the relay assuming that the current through the coil has reached its steady-state value before S is opened.

3.6 The diagram shown below represents a circuit used to determine the speed of a revolver bullet. A and B are strips of tinfoil which are broken by the passage of the bullet as it proceeds from A to B. E is a dc source of potential and V is a reliable voltmeter.

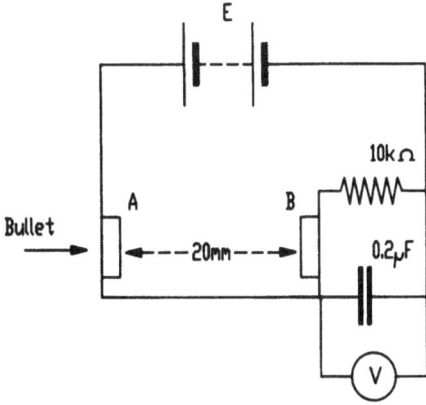

If the voltmeter reads 200 V before the revolver is fired and 190 V after the bullet has broken both strips of tinfoil, calculate the speed of the bullet. You may assume that the resistances of the connecting wires and the tinfoil are negligible.

3.7 A capacitor of capacitance $2\,\mu F$ is charged to a potential of $10\,V$ and is then discharged through a circuit containing a self-inductance of $5\,mH$ and a total resistance R. If the resistance R in the circuit has been adjusted so that critical damping occurs, determine

1. at what time, after the discharge begins, the maximum current occurs, and
2. the value of that maximum discharge current.

4 Alternating current theory

4.1 The waveforms shown in (a) of the figure were recorded on a CRO when the instrument was connected first across R and then

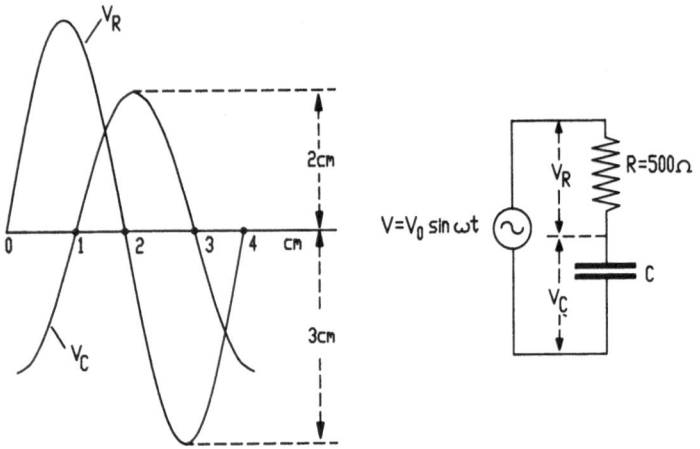

across C in the circuit shown in (b). The horizontal scale in (a) was $5\,ms\,cm^{-1}$ and for both traces the vertical scale was $20\,V\,cm^{-1}$.

From the information given above determine

1. the supply frequency,
2. the rms voltages across R and C,
3. the rms current,
4. the capacitance C,
5. the peak voltage V_0 applied to the circuit.

4.2 A resistor of resistance $400\,\Omega$ and a capacitor of capacitance $0.1\,\mu F$ are connected in parallel and the combination is connected across an ac voltage source. The voltage across the terminals of the ac source is 1 V but, when the parallel circuit is removed, it is observed to rise to 1.4 V. If the internal resistance of the ac source is $50\,\Omega$, calculate the impedance of the parallel combination and the frequency of the source.

4.3 The voltage applied to a series circuit lags behind the current by 40°. If the circuit consists of a resistor of $10\,\Omega$ in series with a reactance, determine the value of the reactance and state whether it is inductive or capacitive. If the frequency of the applied voltage is 1.2 kHz, determine the value of the reactive component.

4.4 A 1 kHz ac source of peak voltage 100 V is connected across a circuit consisting of a capacitor of capacitance $0.5\,\mu F$ in parallel with an inductor of self-inductance 20 mH, which is in turn in parallel with a resistor of resistance $100\,\Omega$. Determine the magnitude and phase of the peak current from the source.

4.5 For the circuit shown in the diagram, show that the reactance between the terminals A and B is zero at all frequencies if $R^2 = L/C$. (Note: R–C in series across a large inductance can be used to suppress the high transient voltages which develop when an ac source is suddenly disconnected.)

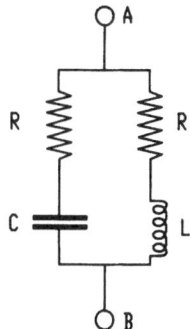

4.6 A network consists of a resistance R and a reactance X connected in series. When an ac voltage of $339\sin(314t)$ V is applied to the network, it is found that the average power is 2.5 kW and the power factor is 0.342 lagging. Determine the values of the two elements in the series circuit.

4.7 A source of frequency 50 Hz with an rms output of 240 V has an internal impedance which is equivalent to a 200 Ω resistor in series with a 0.318 H inductor. Determine the value of resistive load for maximum transfer of power and and calculate the resulting power dissipated in the load (see section 4.16, condition 1).

4.8 A signal generator, operating at a frequency of 5 kHz and having an output resistance of 600 Ω and an rms voltage output of 20 V, has a load which is a variable inductor of inductance L whose resistance is given by $R_L = 1500L\,\Omega\,H^{-1}$. Determine the value of inductance and resistance of the inductor required for maximum power transfer to the load and calculate the power dissipated in the load (see section 4.16, condition 2).

4.9 Determine the generalized impedance $Z(s)$ of the network shown in the diagram. Hence calculate the steady-state values of the currents I_1 and I_2 flowing in the two branches when the following voltage generators are connected across the input terminals AB: (a) a dc generator $V_G = V_0$, and (b) a sinusoidal generator $V_G = V_0 \sin(\omega t + \phi)$.

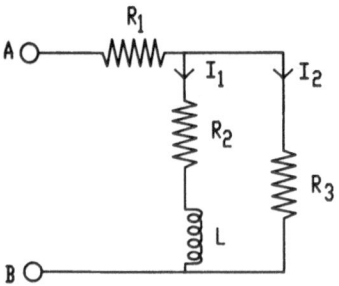

5 Mesh or loop analysis and nodal analysis

5.1 Using either mesh or nodal analysis, determine the current flowing through the 3 Ω resistor in the circuit shown.

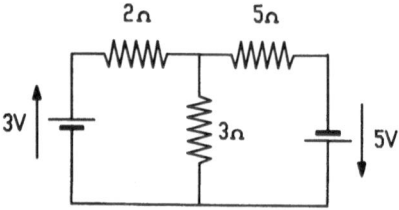

5.2 Using mesh analysis determine the voltage V' as shown in the circuit. Check your result by using nodal analysis.

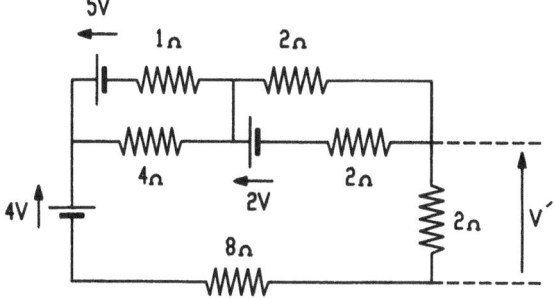

5.3 Determine the pd V' as shown in the circuit by applying nodal analysis to the circuit.

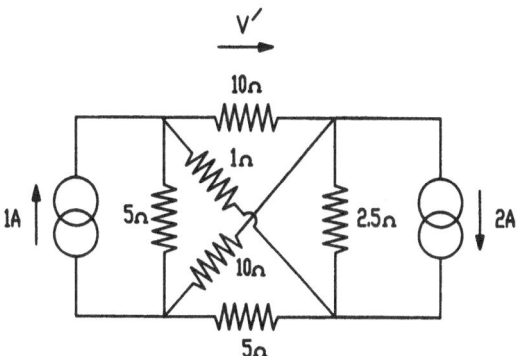

5.4 Determine the current flowing from the $10 \angle 0°$ V generator.

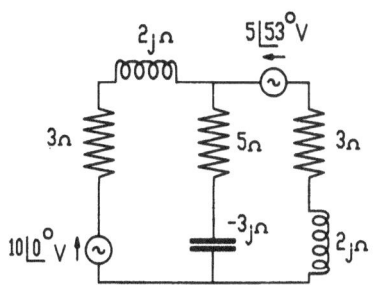

5.5 Determine the pd V' as shown in the circuit by applying mesh analysis to the circuit and check your result by nodal analysis.

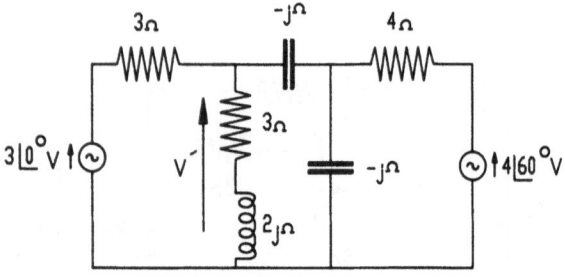

6 Network theorems and transformations

6.1 Determine the Thévenin equivalent voltage generator for the circuit shown in the figure as seen from the terminals AB. (Hint: re-draw the circuit so that each source is in parallel with the voltage across AB.)

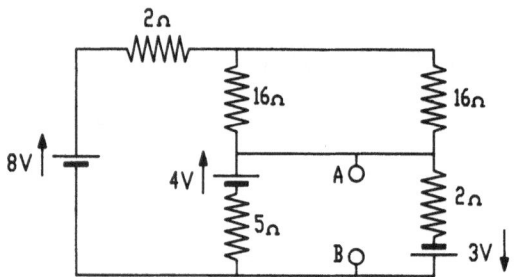

6.2 Determine the Norton equivalent current generator for the circuit shown in the figure with reference to the terminals AB.

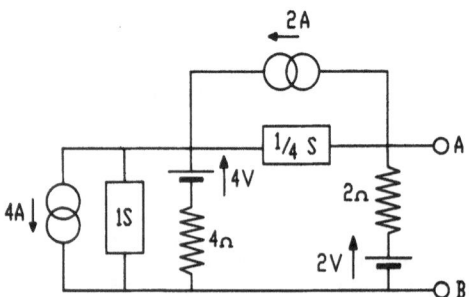

6.3 In the circuit shown in the figure, $V_1 = 5 \sin \omega t$ V and $V_2 = 10 \sin(\omega t + 53°)$ V. Derive, in complex form, Norton's equivalent current generator as seen from the terminals AB.

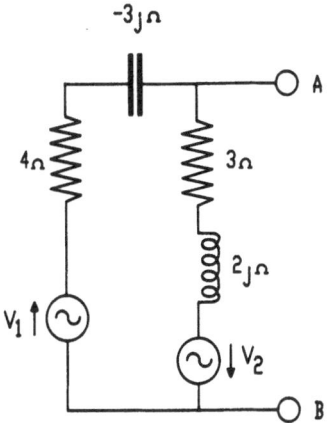

6.4 Determine the Thévenin equivalent voltage generator, in complex form, for the circuit shown in the figure as seen from the terminals AB. What is the complex impedance of the load which when connected across the terminals AB causes maximum power to be produced in the load? Assuming that the voltages and currents are given as rms values, determine the energy dissipated in the above load in 60 s.

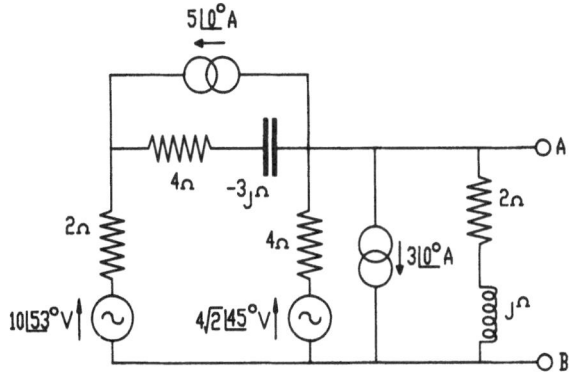

6.5 Calculate the complex emf of the source which can replace the 2 Ω resistor across the terminals AB without changing the currents in the other components of the circuit. Verify that your answer agrees with the statement of the substitution theorem.

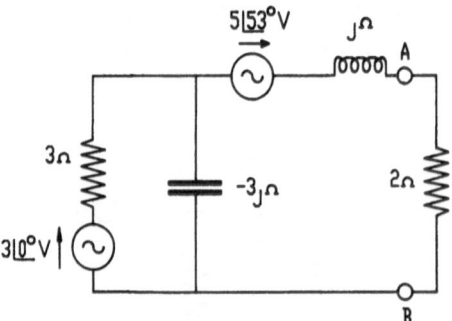

6.6 By using the superposition theorem, calculate the current *I* flowing through the 4 Ω resistor as illustrated in the circuit diagram. Verify your result by using mesh analysis.

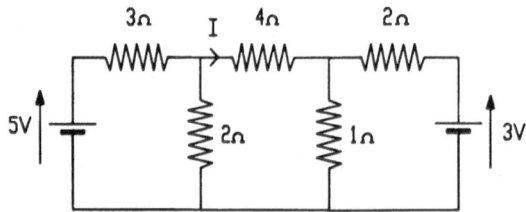

6.7 Using the star–delta transformation determine the input resistance as measured across the terminals AB. Verify your result by use of the definition of the input impedance given by (5.16).

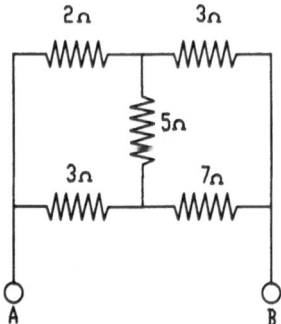

7 Electrical resonance

7.1 It is required to construct a series R–C–L circuit having a resonant frequency of 500 kHz and a linewidth, i.e. the width of the resonance curve at the half-power points, of 25 kHz. If the capacitor to be used in the circuit has a capacitance of 200 pF, determine the values of the inductance L and resistance R required to fulfil these conditions.

7.2 The total resistance R of a series R–C–L circuit consists of the resistance R_L of the inductance coil plus a variable resistance r. At resonance it is found that when $r = 40\,\Omega$ the ratio of the peak voltage across the capacitor to the amplitude of the emf applied to the circuit is 10 and when r is reduced to $10\,\Omega$ the corresponding ratio is increased to 25. If resonance occurs when $\omega_0 = 20 \times 10^3\,\mathrm{rad\,s}^{-1}$, calculate the resistance and self-inductance of the coil and the capacitance of the capacitor required to produce resonance.

7.3 A coil has a self-inductance of 0.5 mH and is tuned to resonance at a frequency of 50 kHz by means of a variable series capacitor. If the Q of the circuit is 30 and the frequency is kept constant, determine the variation in capacitance of the capacitor between the two points on either side of the resonance peak at which the current falls to half of its maximum value.

7.4 A lossy capacitor which can be represented by a capacitance of 0.2 nF in parallel with a resistance of $100\,\mathrm{k\Omega}$ is connected in series with a coil and the combination resonates at an angular frequency of $10^6\,\mathrm{rad\,s}^{-1}$. The half-power linewidth of the response curve is $10^5\,\mathrm{rad\,s}^{-1}$. Determine the inductance and resistance of the coil. (Hint: the solution of this problem can be simplified by representing the lossy capacitor by an equivalent capacitance and series resistor.)

7.5 A circuit consists of a coil of self-inductance L and resistance r which is connected in parallel with a lossy capacitor which may be represented by a capacitance C in parallel with a resistance R. By defining resonance as the condition when the admittance of the circuit is real, show that the resonant frequency of the circuit is

$$f_0 = \frac{1}{2\pi}\left(\frac{1}{LC} - \frac{r^2}{L^2}\right)^{1/2}$$

and that the parallel resistance of the circuit at resonance is given by

$$R_\mathrm{p} = \frac{RL}{L + RCr}$$

By assuming that the Q of the circuit is high show that it is approximately given by

$$\frac{1}{Q} = \frac{1}{R}\sqrt{\frac{L}{C}} + r\sqrt{\frac{C}{L}}$$

(Hint: when determining Q of the circuit use the general definition of Q given by (7.29).)

8 Coupled circuits

8.1

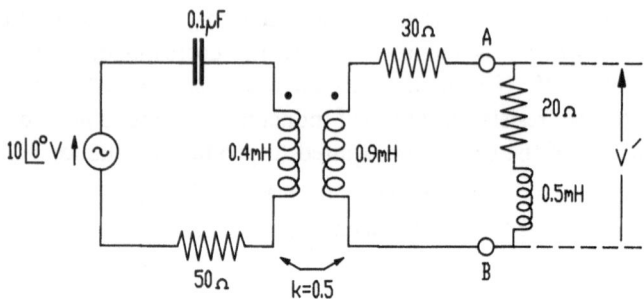

If the angular frequency of the ac generator is 10^5 rad s^{-1}, determine the voltage V' across the coil connected to the terminals AB of the circuit shown in the diagram.

8.2 Derive Thévenin's equivalent voltage generator for the circuit of exercise 8.1 with respect to the terminals AB. Check your result by determining the voltage V' across the coil of resistance $20\,\Omega$ and self-inductance 0.5 mH, as used in exercise 8.1, when it is connected across the output terminals of the equivalent voltage generator.

8.3 Determine the current flowing through the $10\,\Omega$ resistor in the circuit shown in the diagram. Check your result by deriving the appropriate circuit equations by replacing each of the mutual inductances by three self-inductances by using the transformation described in section 6.9.

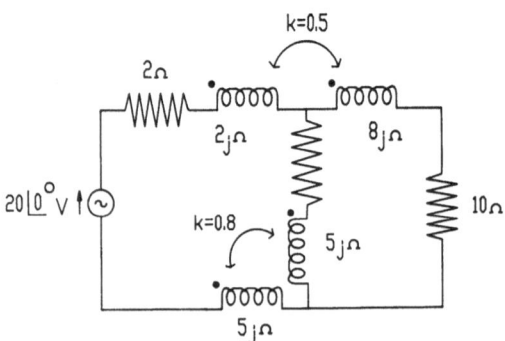

8.4 In the circuit shown in Fig. 8.9(a) on page 96 the primary and secondary circuits are tuned to the same resonant frequency and the circuit components have the following values: $L_1 = 100\,\mu H$, $L_2 = 300\,\mu H$, $C_1 = 150\,pF$, $C_2 = 50\,pF$ and $V_1 = 1\,V$ (rms). If the Q factors of the primary and secondary circuits are 100 and 200 respectively, determine the resonant frequency, and the value of the coupling coefficient required to produce maximum current in the secndary circuit; also determine the value of this current and the corresponding current in the primary circuit.

9 Two-port networks

9.1 Determine the z parameters of the circuit shown in the diagram.

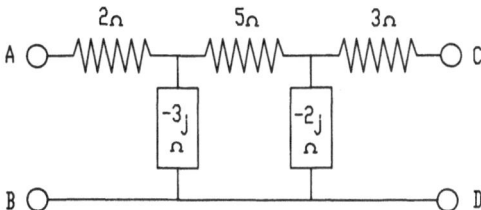

9.2 If a source of emf $V_0 = 3\angle 0°\,V$ and internal impedance $1\,\Omega$ is connected across the terminals AB of the circuit shown in exercise 9.1 and a resistive load of $2\,\Omega$ is connected across the terminals CD determine, using mesh or nodal analysis, the voltage amplification $A = V_2/V_0$ where V_2 is the voltage across the load measured in the direction CD. Determine also the input impedance Z_{IN} of the circuit as measured across the terminals AB. Check both of your answers by directly substituting the values of the z parameters obtained in exercise 9.1 into (9.21) and (9.23).

9.3 Determine the z parameters of the circuit shown in the diagram.

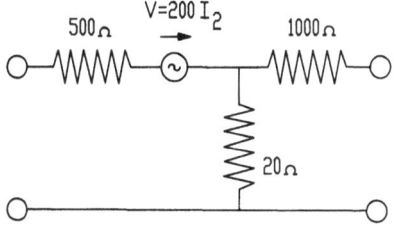

9.4 Determine the *a* parameters of the circuit shown in the diagram and check your result by using (9.27) and (9.28).

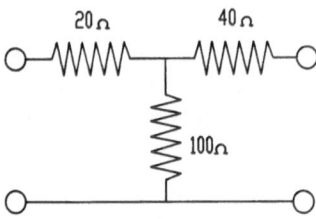

9.5 Determine the image impedances of the circuit shown in exercise 9.4. The insertion loss of a network is defined as $10\log_{10}$ $(P_1/P_2)\,dB$, where P_1 is the power in the load without the network connected between the source and the load and P_2 is the power in the load when the network is connected between the source and the load. Determine, therefore, the insertion loss of the circuit shown in exercise 9.4 when it is inserted between a matched source and a matched load.

9.6 Determine the cut-off frequency of a constant-k low-pass filter which is composed of Π sections having series arms containing a self-inductance of 40.5 mH and shunt arms containing a capacitance of $0.05\,\mu F$. What attenuation, in decibels, is produced by the section at a frequency of 20 kHz? You may assume that the filter is terminated by its characteristic impedance and that the elements are loss-free.

9.7 A T section of a constant-k band-pass filter is shown in the diagram. If $L_1C_1 = L_2C_2$, show that the extent of the pass band is

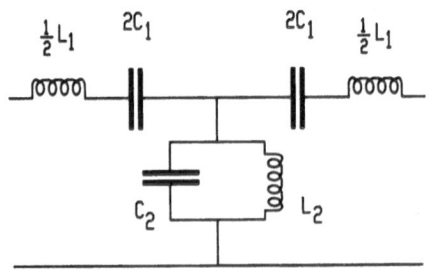

given by the frequencies

$$f_\pm = \frac{k}{2\pi}\left[\left(\frac{1}{L_1 L_2 + L_1^2}\right)^{1/2} \pm \frac{1}{L_1}\right]$$

where $k = (L_2/C_1)^{1/2} = (L_1/C_2)^{1/2}$.

9.8 A T section of a constant-k band-stop filter is shown in the diagram. If $L_1 C_1 = L_2 C_2$, show that the extent of the pass-band is given by the frequencies

$$f_\pm = \frac{k}{4\pi}\left[\left(\frac{4}{L_1 L_2} + \frac{1}{4L_2^2}\right)^{1/2} \pm \frac{1}{2L_2}\right]$$

where $k = (L_1/C_2)^{1/2} = (L_2/C_1)^{1/2}$.

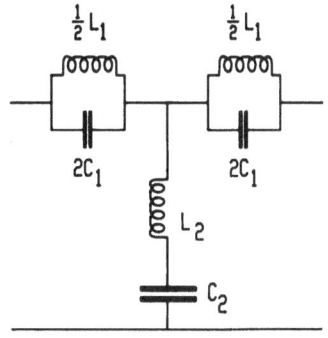

APPENDIX C

Answers to exercises

2 Direct current theory

2.1 1.37 A through the 15 V battery, -0.75 A through the 10 V battery, 7.69 W

2.2 100 V, 20 Ω

2.3 $R = (24 - 0.52n)\Omega$, maximum number of cells is 46; 56.25% wasted energy

3 Capacitors, inductors and transients

3.1 1. 4×10^3 A s^{-1}; 2. 0.5 A; 3. 2×10^3 A s^{-1}; 4. 2.5 A; 5. 0.1 ms

3.2 (a) 2.972 Hz; (b) 59.44 μA

3.3 45.14 A

3.4 371.7 MΩ

3.5 1. 2.16 H; 2. 1.06 ms

3.6 195 m s^{-1}

3.7 1. 10^{-4} s; 2. 73.6 mA

4 Alternating current theory

4.1 1. 50 Hz; 2. $V_{R(rms)} = 42.4$ V, $V_{C(rms)} = 28.3$ V; 3. $I_{rms} = 85$ mA; 4. 9.6 μF; 5. 72.1 V

4.2 125 Ω, 12.09 kHz

4.3 15.81 μF

4.4 1.11 A lagging behind the voltage by 25° 44′

4.6 $R = 2.69$ Ω, $L = 23.5$ mH

4.7 223.6 Ω, 68 W

4.8 19.1 mH, 28.7 Ω, 15 mW

4.9 $Z(s) = R_1 + R_3(R_2 + sL)/(R_2 + R_3 + sL)$

(a) $I_1 = V_0 R_3/(R_1 R_2 + R_2 R_3 + R_3 R_1)$,
$I_2 = V_0 R_2/(R_1 R_2 + R_2 R_3 + R_3 R_1)$

(b) $I_1 = \dfrac{V_0 R_3 \sin(\omega t + \phi - \phi_1)}{[(R_1 R_2 + R_2 R_3 + R_3 R_1)^2 + \omega^2 L^2 (R_1 + R_2)^2]^{1/2}}$

where $\tan \phi_1 = \omega L(R_1 + R_3)/(R_1 R_2 + R_2 R_3 + R_3 R_1)$;

$I_2 = \dfrac{V_0 (R_2^2 + \omega^2 L^2)^{1/2} \sin(\omega t + \phi - \phi_1 + \phi_2)}{[(R_1 R_2 + R_2 R_3 + R_3 R_1)^2 + \omega^2 L^2 (R_1 + R_3)^2]^{1/2}}$

where $\tan \phi_2 = (\omega L/R_2)$.

5 Mesh or loop analysis and nodal analysis

5.1 0.161 A

5.2 $V' = -0.169$ V

5.3 $V' = -3.837$ V

5.4 $1.162 \; \underline{/-33.21°}$ A

5.5 $V' = 1.78 \; \underline{/-8.55°}$ V

6 Network theorems and transformations

6.1 $V_{oc} = 0.125$ V, $R_{AB}^{OUT} = 1.25$ Ω

6.2 $I_{sc} = -1.17$ A, $G_{AB}^{OUT} = 0.71$ S

6.3 $I_{sc} = -(1.815 + 0.325\,j)$ A, $Y_{AB}^{OUT} = (0.391 - 0.034\,j)$ S

6.4 $V_{oc} = -(7.36 - 2.75\,j)$ V, $Z_{AB}^{OUT} = (1.24 + 0.21\,j)$ Ω, $Z_L = 1.24 - 0.21\,j$; 12.45 W

6.5 $V_0 = (2.32 + 1.76\,j)$ V

6.6 0.17 A

6.7 3.32 Ω

7 Electrical resonance

7.1 $L = 0.507$ mH, $R = 79.6$ Ω

7.2 $R_L = 10$ Ω, $L = 25$ mH, $C = 0.1\ \mu$F

7.3 2.348 nF

7.4 $L = 5$ mH, $r = 250.6\,\Omega$

8 Coupled circuits

8.1 $V' = 1.30 \;\underline{/\,139.5°}\;$ V

8.2 $V_{oc} = 30j/(5 - 6j)$ V, $Z_{AB}^{OUT} = (37.39 + 98.85j)\,\Omega$

8.3 $I = 0.551 \;\underline{/\,-122.35°}\;$ A

8.4 $f_0 = 1.30$ MHz, $k_0 = 7.07 \times 10^{-3}$, $I_2 = 50.0$ mA, $I_1 = 61.1$ mA

9 Two-port networks

9.1 $z_{11} = (2.9 - 2.1j)\,\Omega$, $z_{12} = -0.6(1 + j)\,\Omega$, $z_{21} = -0.6(1 + j)\,\Omega$, $z_{22} = (3.4 - 1.6j)\,\Omega$. Note: $z_{12} = z_{21}$ which indicates a passive network.

9.2 $A_v = 0.67 \;\underline{/\,-89.05°}$, $Z_{IN} = 3.68 \;\underline{/\,-37.13°}\;\Omega$

9.3 $z_{11} = 520\,\Omega$, $z_{12} = -180\,\Omega$, $z_{21} = 20\,\Omega$, $z_{22} = 1020\,\Omega$. *Note*: $z_{12} \neq z_{21}$ indicates an active element.

9.4 $a_{11} = 1.2$, $a_{12} = 68\,\Omega$, $a_{21} = 0.01$ S, $a_{22} = 1.4$

9.5 $76.35\,\Omega$, $89.06\,\Omega$, 6.5 dB

9.6 $f_c = 5$ kHz, 15.3 dB

Index

A-parameters 109
Accepter circuit 87
Active component 3
Admittance 38
  generalized 47
  input 59
  output 59, 60
  parameters 108
  transfer 59, 62
Ampere 2, 136
Amplication
  current 114
  voltage 114
Angular frequency 23, 136
Attenuation coefficient 118
Attenuators 128
  ladder network 129

Capacitance 5
Capacitor 5
Capacitors
  in parallel 15
  in series 14
Characteristic impedance 116
Charge 1
Circuit magnification
  factor 83
Coefficient of coupling 5, 94
Compensation theorem 74

Complex-frequency 46
  impedance 32
  representation 32
Connor, F.R. 135
Coulomb 2
Coupled circuits 90
  applications 105
Coupling
  coefficient 94
  direct 103
  impedance 90
Cramer's rule 51
Critical
  coupling 101
  damping 22
Current 1
  generator 8, 9
  magnification factor 86
  response curves 82
  at resonance 98
  near resonance 98

Damping 2
  critical 22
  over 21
  under 21
Delay lines 134
  artificial 134
Delay networks 134

Design resistance 124
Direct coupling 103
Dobbs, E.R. 1
Driving-point impedance 54
Dynamic impedance 84

Eddy current loss 96
Electric field 2
Electromotive force 3
Emf 3

Farad 5
Feeder lines 85
Filters 123
  constant-k 124
  high-pass 125
  low-pass 124
  pass-band 123
  prototype 124
  stop-band 123
Frequency
  resonant 79
  cut-off 121

H-parameters 108
Half-power points 82
Half-section 121
Henry 4
Hybrid parameters 108
Hysteresis loss 96

Impedance 31
  characteristic 117, 131
  complex 32
  distributed 129
  dynamic 84
  generalized 47
  image 121
  input 54, 114
  iterative 117, 122
  lumped 129

output 55, 114
parallel 84
parameters 107
Impedances in parallel 37
  in series 36
Inductance 4
  leakage 95
  mutual 4, 35
  self 5
Inductors in parallel 17
  in series 16
Input admittance 59
Input impedance 54
Joule heating 3

Kirchhoff's laws 6, 36

L-section 121
Ladder networks 122
Leakage inductance 95
Linewidth (of resonance) 81
Loop 6, 48
Loop analysis 48
Losses
  copper 96
  eddy current 96
  hysteresis 96
  iron 96
Low-loss approximation 132

Magnification factor 83
Matched load 13, 44
Maximum power theorem 13
Mesh 6, 48
Mesh analysis 48
  and mutual inductance 52
Millman's theorem 69
Mutual inductance 4, 35, 77

Natural frequency 80
Neper 46, 118

Neper frequency 46
Networks
 cascaded 114
 chain 117
 delay 134
 ladder 117
 symmetrical 118
 2-port 106
Neuman's theorem 4
Nodal analysis 57
Node 6
Non-linear elements 12
Norton's theorem 66

Ohm's law 3
Output admittance 39
Output impedance 55
Over-damping 21

Network 118, 120

Parallel impedance 84
Parallel L–C–R 84
Passive component 3
Parametric representation of 2-
 port network 107
 conversion 110
Phase angle 25, 33, 45
Phase-change coefficient 118
Phase difference 25
Phasor 26
Phasor diagram
 rotating 26
 stationary 27
P–n junction 12
Potential difference 2
Power
 apparent 40
 average 40
 complex 42
 factor 40

instantaneous active 40
instantaneous reactive 40
mean 38
triangle 42
usage 73
Primary current 91
Propagation constant 117
Pulsatance 33

Q 81
Quality factor 81
 definitions of 88

R–C circuit 19
R–C–L circuit 20
Reactance 30
Reciprocity theorem 72
Reflected impedance 91, 94
Rejector circuit 88
R–L circuit 17
Resistance 3
 design 124
 internal 4
Resistors in parallel 11
 in series 10
Resonance 79
Resonance curves 82
Resonance (definitions of) 86
Resonant frequency 79
Root mean square values 38
Rms current 39

S-notation 46
SI units Preface, 136
Self-inductance 5
Siemen 3
Start-delta transformation 74
Substitution theorem 74
Superposition theorem 73
Symbols 136
Symmetrical networks 118

T-network 118
Temperature coefficient of
    resistance 3
Thevenin's theorem
Time constant 18
Transformer
  low-frequency 92
  perfect 93
Transfer admittance 59, 62
Transfer impedance 56
Transients 17
Transmission line 129
  dispersion 133
  distortion 133
Transmission matrix 115

Transmission parameters 109

Under-damping 21

Vacuum diode 12
Var 40
Varistor 12
Volt 2
Voltamperes 40
Voltage generators 8, 9
Voltage magnification 83

Y-parameters 107

Z-parameters 107